FOL

D0061364

TABLE OF CONTENTS

3 CENTROIDS AND MOMENTS OF INERTIA

4 VECTORS

5 KINEMATICS

6 STATICS

7 KINETICS

E-I-T REVIEW
A Study Guide for the Fundamentals of Engineering Exam

Second Edition

The late VIRGIL M. FAIRES
The late JOY O. RICHARDSON

With a preface by Michael R. Lindeburg, P.E.

Professional Publications, Inc.
Belmont, CA 94002

Printed in the United States of America

ISBN: 0-932276-69-5

Professional Publications, Inc.
1250 Fifth Avenue, Belmont, CA 94002

Current printing of this edition: 4

8 STRENGTH OF MATERIALS

9 THERMODYNAMICS

10 FLUID MECHANICS

11 HEAT TRANSFER

12 ELECTRICITY

13 CHEMISTRY

14 PHYSICS

15 ENGINEERING ECONOMY

PREFACE TO THE FIRST EDITION

The objective of this book is to reiterate certain principles and ideas so basic and so commonly applied that once they are mastered you will be assured of a certain percentage of achievement on the E-I-T examination. With the addition of a good browsing review of pertinent texts, you should be in a position to do a commendable job. Some of the problems call for knowledge not mentioned or detailed in this book; these will help direct your general review.

Since the examinations can virtually cover the entire undergraduate program of an engineering student, it is evident that the purpose of this book would be defeated if an attempt was made to review everything that might be included on a particular examination. In the solution of problems, rely on the most fundamental principles. Do not be satisfied with substituting in a derived formula unless it is a very basic one that summarizes briefly a much longer theory, such as $s = Mc/I$.

While one might read most of the problems in this book and ponder them for a method of solution, it is unlikely (and unnecessary) that any one student will solve them all. However, the large number of problems is advantageous in spotting principles that, at least in the past, have undergone a certain repetition. In this connection, one should note that the various examinations are undergoing a rapid evolution, and history in this respect will not necessarily repeat itself.

V.M.F.
J.O.R.

PREFACE TO THE FIRST EDITION

[illegible]

[illegible]

[illegible]

PREFACE TO THE SECOND EDITION

We all tend to look back with fondness at events, activities, and items that are part of our pasts. That was my reaction when I learned that the first edition of the *E-I-T Review* by Virgil M. Faires and Joy O. Richardson had gone out of print, and that the original publisher had decided not to bring out a second edition, because this is the book I had used for my own E-I-T examination.

Therefore, I contacted the original publisher, Prentice-Hall, Inc., and requested permission to update and reprint the book. Professional Publications, Inc., has edited, reformatted, and typeset the entire book to create this second edition.

I took the E-I-T examination in California the very first time California used the National Council of Examiners for Engineering and Surveying (NCEES) examination. I had studied for six months, working problems from this book every evening for several hours. And, I passed the examination the first time I took it.

Since then, the NCEES examination has continued to evolve in complexity, coverage, and grading. Because of that, I have removed or changed the emphasis on certain subjects during the creation of the second edition. Of course, there are examination subjects that could have been added, but to do so would have been a duplication of the efforts that went into the *Engineer-In-Training Review Manual*, my first book on the subject.

All of the problems and nearly all of the worked-out examples in this book have been taken from E-I-T examinations given in prior years. This book still contains the wealth of practice problems that helped me through the E-I-T examination. Those problems, specifically oriented to E-I-T examination subjects and at the proper level of complexity, are what make this book so valuable.

I recommend this book as a secondary reference (after my own *Engineer-In-Training Reference Manual*) and a source of additional study problems for any engineer preparing for an E-I-T examination. Perhaps, when you have gone on to bigger and better things in your engineering career, you will be able to look back with fondness at two books that helped you through the E-I-T examination!

Michael R. Lindeburg, P.E.
Belmont, California
August 1987

ACKNOWLEDGMENTS

Thank you, David S. Goldstein and Don Wilkins, for proofreading the original manuscript, galleys, and page proofs, and also for improving and updating this book from both an English and engineering standpoint.

Thank you, Rhonda Jones and Melissa Westover, for coordinating production of this book. You gave the readers a polished finished product.

Thank you, Yasuko Kitajima and Richard Weyhrauch of Aldine Press, for typesetting another Professional Publications book. With your technical knowledge and familiarity with our style, you were not afraid to make corrections to the manuscript.

A special thank you goes to Prentice-Hall, Inc., without whose assistance this reprint of one of their titles would not have been possible.

Thanks to you all!

Michael R. Lindeburg, P.E.
President
Professional Publications, Inc.

ABBREVIATIONS IN COMMON USAGE

AC	alternating current
Atm	atmospheres, a unit of pressure (14.7 psi)
BHN	Brinell hardness number
BTU	British thermal unit
CCW	counterclockwise
cc	cubic centimeter
cfm	cubic feet per minute
CW	clockwise
cm	centimeter
cpm	cycles per minute
cps	cycles per second
DC	direct current
g	gram
gpm (gal/min)	gallons per minute
hp	horsepower
ID	inside diameter
k, K	kips
kw	kilowatt
kw-hr	kilowatt-hour
ln	natural logarithm
log	logarithm to base 10
mm	millimeters
mph	miles per hour
OD	outside diameter
psf	pounds per square foot
psi	pounds per square inch
psia	pounds per square inch absolute
psig	pounds per square inch gauge
rpm	revolutions per minute
rps	revolutions per second

1 *INTRODUCTION*

This book is primarily a guide to preparation for the Fundamentals of Engineering (F.E.) examination, also known as the Engineer-in-Training (E-I-T) examination. It generally says what *is*, not *why*; to review the "why's," refer to your basic texts. The statements are necessarily brief, often to the point of not being rigorous. The assumption is that detailed studies of the various subjects have already been made and that this outline needs only to *remind* you. When the meaning and significance of the statements are not readily recalled, the only satisfactory recourse is to refer to a more complete work, which should always be near at hand as you start the review.

ORGANIZATION

For explanation purposes, it is desirable to classify the subjects and treat one subject at a time. Thus, there are problems on motion under our heading of kinematics, but actual engineering problems, and some of the problems on the examination, are necessarily broader.

Engineering is the art of applying mathematics and the physical sciences to further the well-being of human-kind. In a single engineering problem, the knowledge of geometry, algebra, analytic geometry, calculus, kinematics, mechanics, thermodynamics, and perhaps other sciences may be necessary. Hence, during this review you should do your best to perceive the interrelations of the various phases of science, looking for analogous relations (for example, heat transfer and the flow of electricity). Moreover, there are often no sharp dividing lines between subjects. Chemistry merges into thermodynamics and vice versa. Is the determination of the amount of water condensed after the products of combustion have been cooled a chemical problem or a thermodynamic one? It doesn't matter. The engineer must find a practical answer.

SYMBOLS

Even though symbols are standardized to a large extent, there are not as many letters available as there are ideas to be symbolized. Thus, the same letter often represents more than one concept. For this reason, it is not always convenient to use a standard symbol. Moreover, there is no reason to suppose that standard symbols will be used in all questions of all examinations. Consequently, it is important for you to learn a principle *in words.* (Meaningful thinking is done in words. The more words one commands, the better the possible grade of thinking.) Symbols are the engineer's shorthand for words. Be prepared to use the principles even though the symbols on the examination are different from those in this book. Actually, most of the problems on the examinations are easy provided you *recognize* and *understand* the principle or principles involved. Our symbols are either defined at the beginning of each chapter, shown in an illustration, or locally defined.

REGISTRATION LAWS

Although there is a so-called "model law" after which most state laws are patterned, the differences from state to state are sufficient that you should communicate with your own state board for detailed information, if you have not already done so.

NATURE OF THE EXAMINATION

The nature (content, emphasis, grading, etc.) of the F.E. examination changes regularly. The examination will be described only briefly here.*

*More information about the F.E. examination is given in the *Engineer-In-Training Reference Manual*, written by Michael R. Lindeburg, also available from Professional Publications, Inc.

The F.E. examination is eight hours long. It covers all engineering subjects normally found in a four-year bachelor of science degree program, as well as mathematics and some science-related subjects from the fields of physics and chemistry. The examination is entirely multiple choice. (Check with your State Board of Registration to make sure, however.) Grading is done by computer, and there is no penalty for guessing.

The examination is administered twice a year, usually in mid-April and late October. There is a fee for taking the examination, and the filing deadline is two to four months prior to the examination. For information about experience and education prerequisites, contact your state's Board of Registration for Professional Engineers.

2 *MATHEMATICS*

GEOMETRY AND TRIGONOMETRY

CIRCLE

$$\text{circumference} = 2\pi r = \pi D$$

$$\text{area, } A = \pi r^2 = \frac{\pi D^2}{4} = 0.7854D^2$$

PLANE TRIANGLES (Figure 2.1)

The sum of the internal angles is

$$\alpha + \beta + \gamma = 180^\circ = \pi \text{ radians}$$

$$\text{area } A = \frac{1}{2}bh = [s(s-a)(s-b)(s-c)]^{1/2}$$

$$s = \frac{1}{2}(a+b+c)$$

The *median* is the line from a vertex to midpoint of the point side (Aa, Bb, and Cc in figure 2.1(c)).

The *centroid* (also called *center of gravity*) is at the intersection of medians, G.

$$bG = \frac{1}{3}bB$$

$$aG = \frac{1}{3}aA$$

$$cG = \frac{1}{3}cC$$

Any triangle constructed in a semicircle (diameter is one side) is a right triangle.

Law of Sines

$$\frac{a}{\sin\alpha} = \frac{b}{\sin\beta} = \frac{c}{\sin\gamma}$$

Law of Cosines

$$c^2 = a^2 + b^2 - 2ab\ \cos\gamma$$

$$b^2 = a^2 + c^2 - 2ac\ \cos\beta$$

$$a^2 = b^2 + c^2 - 2bc\ \cos\alpha$$

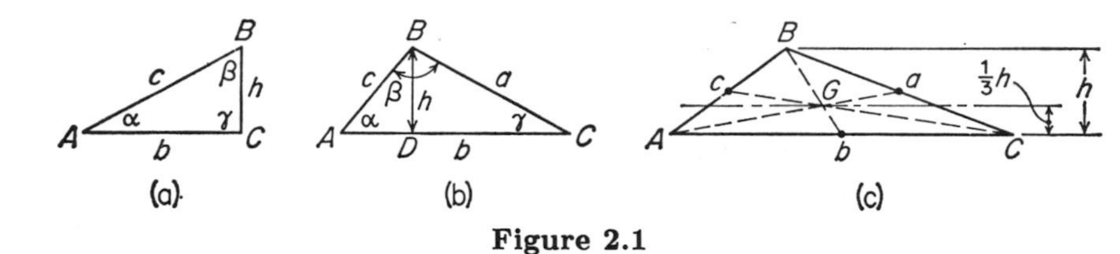

Figure 2.1

Figure 2.1(b): Triangles ADB and DBC are similar (sides mutually perpendicular, $\beta = 90°$). Therefore,

$$\frac{AD}{h} = \frac{h}{DC}$$

$$h^2 = (AD)(DC)$$

Right Triangles

Figure 2.1(a): $c^2 = h^2 + b^2$.

$$\sin\alpha = \frac{h}{c}$$

$$\sin\beta = \frac{b}{c}$$

$$\cos\alpha = \frac{b}{c}$$

$$\tan\alpha = \frac{h}{b}$$

$$\tan\beta = \frac{b}{h}$$

$$\cos\beta = \frac{h}{c}$$

$$\csc\ \alpha = \frac{1}{\sin\alpha}$$

$$\sec\ \alpha = \frac{1}{\cos\alpha}$$

$$\cot\ \alpha = \frac{1}{\tan\alpha}$$

$$\alpha = \text{arc}\sin\frac{h}{c} = \sin^{-1}\frac{h}{c} = \text{arc}\cos\frac{b}{c}$$

$$= \cos^{-1}\frac{b}{c} = \text{arc}\tan\frac{h}{b} = \tan^{-1}\frac{h}{b}$$

Arc sin means *the angle whose sine is*; similarly for *arc cos* and *arc tan.*

Often considerable time is saved by knowing the values of the following trigonometric functions:

$$\sin 0° = \cos 90° = 0$$

$$\sin 90° = \cos 0° = 1$$

$$\tan 0° = 0$$

$$\tan 45° = 1$$

$$\tan 90° = \infty$$

$$\sin 30° = \cos 60° = 0.5$$

$$\sin 60° = \cos 30° = 0.866 = \frac{1}{2}\sqrt{3}$$

$$\sin 45° = \cos 45° = 0.707 = \frac{1}{2}\sqrt{2}$$

Radians

The angle subtended by an arc on the circumference of a circle equal in length to the radius of the circle is a *radian*. Since there are 2π radians in a circle, there are 2π radians in 360°.

$$1 \text{ radian} = \frac{360^\circ}{2\pi} \approx 57.3^\circ$$

An angle appearing in an equations, as $s = r\theta$, is in radians, unless there has been an intentional conversion of units. Any unit other than radians in such a situation should be clearly stated.

Angles Greater Than 90°

Figure 2.2(b):

$$\begin{aligned} \sin\theta &= \sin(180^\circ - \delta) = \sin\delta \\ \cos\theta &= \cos(180^\circ - \delta) = -\cos\delta \\ \tan\theta &= \tan(180^\circ - \delta) = -\tan\delta \end{aligned}$$

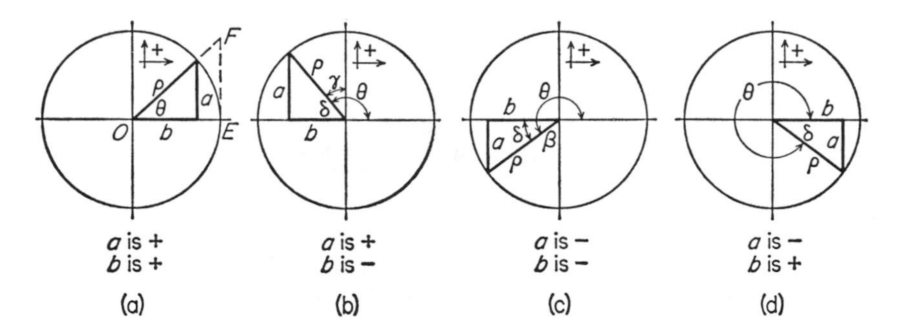

Figure 2.2 In each illustration, $\sin\theta = a/\rho$, $\cos\theta = b/\rho$, and $\tan\theta = a/b$, with proper accounting of negative values of a and b, as in (c), where both a and b are negative. In (a), $\tan\theta = FE/OE = FE$, when $OE = \rho = 1$; similarly in the other illustrations.

Figure 2.2(c):

$$\begin{aligned} \sin\theta &= -\sin(\theta - 180^\circ) = -\sin\delta \\ \cos\theta &= -\cos(\theta - 180^\circ) = -\cos\delta \\ \tan\theta &= +\tan(\theta - 180^\circ) = \tan\delta \end{aligned}$$

Figure 2.2(d):

$$\sin\theta = \sin(360^\circ - \delta) = -\sin\delta$$
$$\cos\theta = \cos(360^\circ - \delta) = +\sin\delta$$
$$\tan\theta = \tan(360^\circ - \delta) = -\tan\delta$$

FUNCTIONS OF ANGLES

$$\sin^2\alpha + \cos^2\alpha = 1$$
$$\sin\alpha = \sqrt{1 - \cos^2\alpha}$$
$$\tan\alpha = \frac{\sin\alpha}{\cos\alpha}$$
$$1 + \tan^2\alpha = \frac{1}{\cos^2\alpha}$$
$$\sin(\alpha \pm \beta) = \sin\alpha\cos\beta \pm \cos\alpha\sin\beta$$
$$\cos(\alpha \pm \beta) = \cos\alpha\cos\beta \mp \sin\alpha\sin\beta$$
$$\tan(\alpha \pm \beta) = \frac{\tan\alpha \pm \tan\beta}{1 \mp \tan\alpha\tan\beta}$$
$$\sin 2\theta = 2\sin\theta\cos\theta$$
$$\cos 2\theta = \cos^2\theta - \sin^2\theta = 2\cos^2\theta - 1 = 1 - 2\sin^2\theta$$
$$\tan 2\theta = \frac{2\tan\theta}{1 - \tan^2\theta}$$
$$\tan 2\theta = \frac{\sin 2\theta}{\cos 2\theta}$$
$$\sin\frac{\theta}{2} = \sqrt{\frac{1 - \cos\theta}{2}}$$
$$\sin\theta = \sqrt{\frac{1 - \cos 2\theta}{2}}$$
$$\sin^2\theta = \frac{1 - \cos 2\theta}{2}$$
$$\cos\frac{\theta}{2} = \sqrt{\frac{1 + \cos\theta}{2}}$$
$$\cos\theta = \sqrt{\frac{1 + \cos 2\theta}{2}}$$
$$\cos^2\theta = \frac{1 + \cos 2\theta}{2}$$

Conversion Constants

LINEAR MEASURE

$12\frac{\text{in}}{\text{ft}}$ $3\frac{\text{ft}}{\text{yd}}$ $5280\frac{\text{ft}}{\text{mi}}$ $80\frac{\text{chains}}{\text{mile}}$ $5.5\frac{\text{yd}}{\text{rod}}$

AREA MEASURE

$144\frac{\text{in}^2}{\text{ft}^2}$ $9\frac{\text{ft}^2}{\text{yd}^2}$ $640\frac{\text{acres}}{\text{mi}^2}$ $10\frac{\text{chains}^2}{\text{acre}}$

VOLUME MEASURE

$1728\frac{\text{in}^3}{\text{ft}^3}$ $7.48\frac{\text{gal}}{\text{ft}^3}$ $4\frac{\text{qt}}{\text{gal}}$ $8\frac{\text{pt}}{\text{gal}}$ $43{,}560\frac{\text{ft}^3}{\text{acre-ft}}$

TIME MEASURE

$60\frac{\text{sec}}{\text{min}}$ $60\frac{\text{min}}{\text{hr}}$ $3600\frac{\text{sec}}{\text{hr}}$ $24\frac{\text{hr}}{\text{day}}$ $365\frac{\text{day}}{\text{yr}}$

PRESSURE MEASURE

$144\frac{\text{psi}}{\text{psf}}$ $0.49\frac{\text{psi}}{\text{in Hg}(60°\text{F})}$ $0.0361\frac{\text{psi}}{\text{in H}_2\text{O}(60°\text{F})}$ $29.92\frac{\text{in Hg}(32°\text{F})}{\text{atm}}$

FORCE AND MASS MEASURE

$16\frac{\text{oz}}{\text{lbf}}$ $1000\frac{\text{lbf}}{\text{kip}}$ $2000\frac{\text{lbf}}{\text{ton}}$ $7000\frac{\text{grains}}{\text{lbm}}$ $32.17\frac{\text{lbm}}{\text{slug}}$

ENERGY AND POWER MEASURE

$778\frac{\text{ft-lbf}}{\text{BTU}}$ $33{,}000\frac{\text{ft-lbf}}{\text{hp-min}}$ $550\frac{\text{ft-lbf}}{\text{hp-sec}}$ $3412\frac{\text{BTU}}{\text{kw-hr}}$ $1.341\frac{\text{hp}}{\text{kw}}$

EXAMPLE: Convert 60 mph to feet per second.

SOLUTION:

$$\left(60\frac{\text{mi}}{\text{hr}}\right)\left(5280\frac{\text{ft}}{\text{mi}}\right)\left(\frac{\text{hr}}{3600\text{ sec}}\right) = 88\text{ ft/sec}$$

ALGEBRA

Equations representing physical laws must be homogeneous; i.e., each term must have the same units. The way to a solution to a problem is often found from the knowledge that units must be consistent.

FUNDAMENTAL OPERATIONS

The roots of the quadratic equation $ax^2 + bx + c = 0$ are

$$x = \frac{-b \pm \sqrt{b^2 - 4ac}}{2a} \qquad \text{[QUADRATIC FORMULA]}$$

Generally, the quickest way to solve cubic and higher-power equations is by trial-and-error until the equation balances.

Any positive number a greater than 1 raised to any positive power $n > 1$ is greater than a.

$$a^n > a \quad \text{when} \quad a > 1 \quad \text{and} \quad n > 1 \quad (\text{e.g.}, 2^2 = 4)$$

Similarly,

$$a^n < a \quad \text{when} \quad a > 1 \quad \text{and} \quad n < 1 \quad (\text{e.g.}, 4^{0.5} = 2)$$
$$a^n < a \quad \text{when} \quad a < 1 \quad \text{and} \quad n > 1 \quad (\text{e.g.}, 0.2^2 = 0.04)$$
$$a^n > a \quad \text{when} \quad a < 1 \quad \text{and} \quad n < 1 \quad (\text{e.g.}, 0.25^{0.5} = 0.5)$$

Be sure you recognize the following forms. (a is positive. Any number with a zero exponent is equal to unity.)

$$a^0 = 1$$
$$a^{-1} = \frac{1}{a}$$
$$a^3 = \frac{1}{a^{-3}}$$
$$a^{1/3} = \sqrt[3]{a}$$
$$a^{-1/3} = \frac{1}{\sqrt[3]{a}}$$
$$a^{2/3} = \sqrt[3]{a^2}$$
$$\left(\frac{a}{b}\right)^n = \frac{a^n}{b^n}$$
$$a^n b^n c^n = (abc)^n$$
$$(-a)^n = -a^n \quad \text{if } n \text{ is odd} \quad [\text{e.g., } (-2)^3 = -2^3 = -8]$$
$$(-a)^n = a^n \quad \text{if } n \text{ is even} \quad [\text{e.g., } (-2)^2 = 2^2 = 4]$$
$$\sqrt[n]{-a} = -\sqrt[n]{a} \quad \text{if } n \text{ is odd} \quad [\text{e.g., } \sqrt[3]{-8} = -\sqrt[3]{8} = -2]$$
$$\sqrt[n]{-a} = \sqrt[n]{a}\sqrt{-1} \quad \text{(an imaginary number) if } n \text{ is even}$$

When the same number (base) appears in a product with different exponents, *add exponents.*

$$a^n a^m a^p = a^{n+m+p}$$
$$a^n a^{1/m} = a^{n+\frac{1}{m}}$$
$$a^{1/n} a^{-m} = a^{\frac{1}{n}-m} = \frac{a^{1/n}}{a^m}$$

Subtract exponents in a quotient: $\dfrac{a^n}{a^m} = a^{n-m} = \dfrac{1}{a^{m-n}}$

EXAMPLE:

$$(9x^4y)(2x^2y^3) = 18x^6y^4$$
$$\frac{3x^2}{4x^3} = \frac{3}{4}\left(\frac{1}{x}\right) = \frac{3}{4x} = \frac{0.75}{x} = 0.75x^{-1}$$

For the power of a power, *multiply exponents*:

$$(a^n)^m = a^{nm}$$
$$(a^n)^{1/n} = \sqrt[n]{a^n} = a$$
$$(a^n)^{1/m} = a^{n/m} = \sqrt[m]{a^n}$$

a^n is a *power*; a is the *base*; n is the *exponent.*

DETERMINANTS

The value of determinants is obtained as in the following examples:

Second order:

$$\begin{vmatrix} a_1 b_1 \\ a_2 b_2 \end{vmatrix} = a_1 b_2 - a_2 b_1$$

Third order:

$$\begin{vmatrix} a_1 b_1 c_1 \\ a_2 b_2 c_2 \\ a_3 b_3 c_3 \end{vmatrix} = a_1 \begin{vmatrix} b_2 c_2 \\ b_3 c_3 \end{vmatrix} - a_2 \begin{vmatrix} b_1 c_1 \\ b_3 c_3 \end{vmatrix} + a_3 \begin{vmatrix} b_1 c_1 \\ b_2 c_2 \end{vmatrix}$$

$$= a_1(b_2 c_3 - b_3 c_2) - a_2(b_1 c_3 - b_3 c_1) + a_3(b_1 c_2 - b_2 c_1)$$

Fourth order:

$$\begin{vmatrix} a_1 b_1 c_1 d_1 \\ a_2 b_2 c_2 d_2 \\ a_3 b_3 c_3 d_3 \\ a_4 b_4 c_4 d_4 \end{vmatrix} = a_1 \begin{vmatrix} b_2 c_2 d_2 \\ b_3 c_3 d_3 \\ b_4 c_4 d_4 \end{vmatrix} - a_2 \begin{vmatrix} b_1 c_1 d_1 \\ b_3 c_3 d_3 \\ b_4 c_4 d_4 \end{vmatrix} + a_3 \begin{vmatrix} b_1 c_1 d_1 \\ b_2 c_2 d_2 \\ b_4 c_4 d_4 \end{vmatrix} - a_4 \begin{vmatrix} b_1 c_1 d_1 \\ b_2 c_2 d_2 \\ b_3 c_3 d_3 \end{vmatrix}$$

The term (A) is called the *minor* of a_1; term (B) is the minor of a_2; etc. Thus, to evaluate an nth order determinant, take the elements of the first column and multiply each by its minor, letting the first product be positive, the next negative, etc., and alternating signs with each product.

LOGARITHMS

Exponents are logarithms. In $a^n = b$, the exponent n is the logarithm of b to the base a:

$$\log_a b = n$$

For the base 10, the abbreviation is generally $\log N$. The 10 is understood.

$$10^{-1} = \frac{1}{10} = 0.1$$

$$10^0 = 1$$

$$10^1 = 10$$

$$10^2 = 100$$

$$\log_{10} 0.1 = -1$$

$$\log_{10} 1 = 0$$

$$\log_{10} 10 = 1$$

$$\log_{10} 100 = 2$$

Natural or Naperian logarithms have a base of $e = 2.71828\ldots$. They are abbreviated $\log_e N$ and $\ln N$.

$$e^0 = 1$$
$$e^1 = e$$
$$e^2 = (2.718\ldots)^2$$
$$e^{-1} = \frac{1}{e}$$
$$\ln 1 = 0$$
$$\log_e e = \ln e = 1$$
$$\log_e e^2 = \ln e^2 = 2$$
$$\log_e e^{-1} = \ln \frac{1}{e} = -1$$

The logarithm of any number less than 1 is negative.

The logarithm of any number greater than 1 is positive.

The logarithm of a number to the base 10 between 10 and 100 is between 1 and 2.

The logarithm of a number to the base e between e and e^2 is between 1 and 2.

To multiply two numbers, add their logarithms, and find the antilog. This is analogous to exponents.

To get ab: Use $\log a + \log b = \log ab = N$, and find the antilog $N = ab$.

For division, subtract logs:

To get $\frac{a}{b}$: Use $\log a - \log b = \log \frac{a}{b} = N$, and find the antilog of N.

$$\log \frac{a}{b} = -\log \frac{b}{a} = \log a - \log b = -(\log b - \log a)$$

If $a^x = N$, then $\log a^x = x \log a = \log N = Q$, and the antilog of Q is the value of $N = a^x$.

Converting logs:

$$\log_e N = 2.30259 \log_{10} N$$
$$\ln N \approx 2.3 \log N$$

If 2.699 is a base-10 logarithm, the digits after the decimal point (0.699) constitute the *mantissa*, and those preceding the decimal (2) constitute the *characteristic*. As taken from a table, the mantissa is always positive, although the logarithm can be negative.

For a number greater than 1, the characteristic is *positive* and always *one less than* the number of digits preceding the decimal.

For a number less than 1, the characteristic is *negative* and always *one greater than* the number of zeros immediately following the decimal point.

If the logarithm is given, look up the number in the table according to the mantissa. For 2.699, from the mantissa 0.699 in a table, find 5. The characteristic 2 indicates that you should add two zeros, which makes the number 500 ($\log 500 = 2.699$).

There are two common ways of indicating the negative logarithm to the base 10:

$$\overline{2}.699, \quad \text{and} \quad 8.699 - 10$$

In any case, the log is $-2 + 0.699 = -1.301$ and would be so used if the log was a part of a product, as 6 log 0.05, etc. However, in looking up the antilog, the form of the log must be such that the mantissa is positive. The antilog $\overline{2}.699$ is 0.05 ($\log 0.05 = \overline{2}.699$).

ANALYTICAL GEOMETRY

STRAIGHT LINES

The equation of a straight line can be stated in several forms, as is convenient. Refer to figure 2.3, which shows a line joining points 1 and 2.

general form: $Ax + By + C = 0$

point-slope form: $y - y_1 = m(x - x_1)$

slope-intercept form: $y = mx + b$

intercept form: $\frac{x}{a} + \frac{y}{b} = 1$

The slope m is

$$m = \tan\theta = \frac{y_2 - y_1}{x_2 - x_1} = \frac{y_2 - y}{x_2 - x} = \frac{y - y_1}{x - x_1}, \text{ etc.}$$

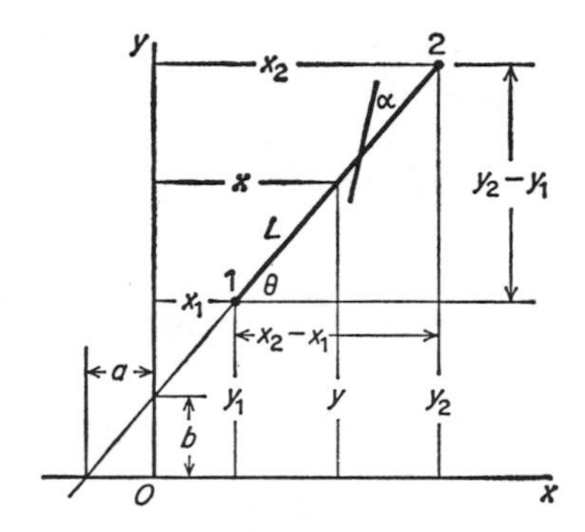

Figure 2.3 As shown, a is negative.

The distance between points 1 and 2 is

$$L = \sqrt{(x_2 - x_1)^2 + (y_2 - y_1)^2}$$

The acute angle α between two straight lines of slopes m_1 and m_2, where $m_2 > m_1$, is such that

$$\tan\alpha = \frac{m_2 - m_1}{1 + m_1 m_2}$$

If two lines are parallel,

$$m_1 = m_2$$

If two lines are perpendicular,

$$m_1 = -\frac{1}{m_2}$$

CIRCLE

The equation of a circle whose center has coordinates (h, k) is

$$(x' - h)^2 + (y' - k)^2 = r^2$$

See figure 2.4. r = radius, and x' and y' are coordinates of any point P on the circle. If the reference axes pass through the center of the circle (x, y axes in figure 2.4; $h = 0$, $k = 0$), the equation reduces to

$$x^2 + y^2 = r^2$$

In parametric form, the equation is

$$x = r\cos\theta$$

$$y = r\sin\theta$$

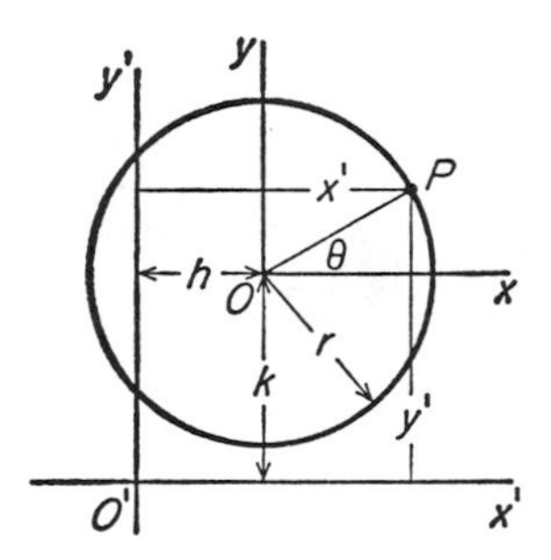

Figure 2.4 Circle

PARABOLA

A parabola is traced by a point P (figure 2.5), whose distances from a fixed point F (called the *focus*) and a fixed line MM (called the *directrix*) are always the same. Referred to $x'y'$ axes (with $O'x'$ parallel to principal axis Ox), where (h, k) are the coordinates of the vertex O,

$$(y' - k)^2 = a(x' - h) = 2p(x' - h)$$

Usually, the equation is referred to axes through the vertex:

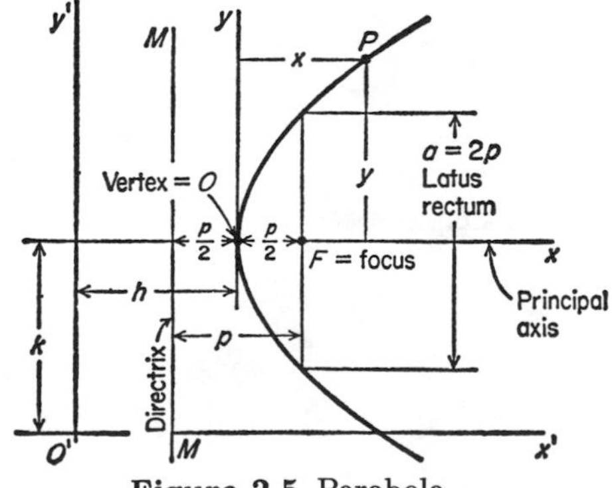

Figure 2.5 Parabola

principal axis, horizontal: $y^2 = ax = 2px$

principal axis, vertical: $x^2 = ay = 2py$

p is the distance between the focus and the directrix.

ELLIPSE

The major axis is $AB = 2a$. The minor axis is $CD = 2b$. For coordinate axes parallel to major and minor axes, the equation of an ellipse (h and k as before) is

$$\frac{(x' - h)^2}{a^2} + \frac{(y' - k)^2}{b^2} = 1$$

With reference to major and minor axes (h and k zero):

$$\frac{x^2}{a^2} + \frac{y^2}{b^2} = 1$$

The foci are F_1 and F_2. A property of the ellipse is that, for all points P on the circumference, $F_1P + PF_2$ is the same.

$$\text{area} = \pi ab$$

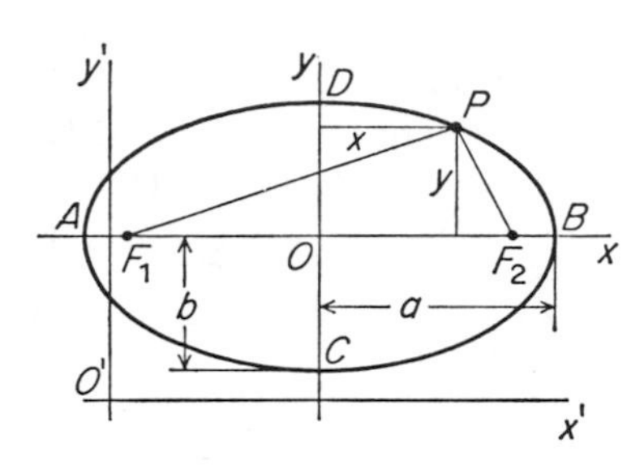

Figure 2.6 Ellipse

HYPERBOLA

For the x axis as the principal axis, solid curves A in figure 2.7,

$$\frac{x^2}{a^2} - \frac{y^2}{b^2} = 1$$

The asymptotes are

$$ay + bx = 0$$
$$ay - bx = 0$$

The positive slope of an asymptote is

$$\tan\theta = b/a$$

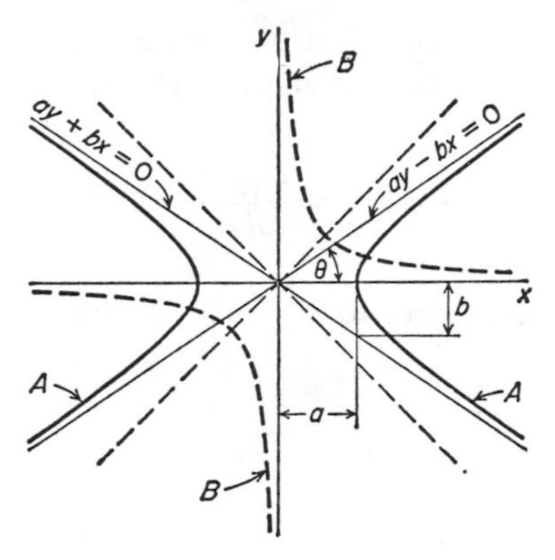

Figure 2.7 Hyperbolas

For equilateral hyperbolas (asymptotes perpendicular), with the x axis as the principal axis, and with $a = b$,

$$x^2 - y^2 = a^2$$

If the x and y axes are the asymptotes, the equation is

$$xy = C = a^2/2$$

See the dotted curves B in figure 2.7. a is measured as shown for the solid curves. A hyperbola has a focus, a directrix, and a latus rectum.

The foregoing curves (circle, parabola, ellipse, and hyperbola) are called *conics* because each of these curves can be obtained from the intersection of a plane and a right cone.

CALCULUS

DIFFERENTIAL CALCULUS

The symbol dx, a derivative, can represent either an infinitesimal change *in* x or an infinitesimal quantity *of* x. The context and the nature of x usually make the meaning clear. Symbols used to mean "function of x" include $\mathrm{f}(x)$, $\mathrm{F}(x)$, $\phi(x)$, and $\psi(x)$.

The value of the function when x is equal to a particular number (say, 2) can be written

$$\mathrm{f}(2), \quad \text{or} \quad \lim_{x\to 2} \mathrm{f}(x)$$

This is read, "limit of the function of x as x approaches 2." The first derivative of the function gives the slope of the curve.

$$\text{slope} = \frac{dy}{dx} = \frac{d\,\mathrm{f}(x)}{dx} = \mathrm{f}'(x)$$

Derivatives can be with respect to another variable (say time, t).

$$\frac{dy}{dt} = \frac{d\,\mathrm{f}(x)}{dt}$$

A derivative of a first derivative is called a *second derivative*, and is written

$$\frac{d(dy/dx)}{dx} = \frac{d^2y}{dx^2} = f''(x)$$

A derivative of the second derivative is called the *third derivative*, and is designated

$$d^3y/dx^3,$$

or

$$f'''(x).$$

If $y = \mathrm{f}(x)$, then

$$y + \Delta y = \mathrm{f}(x + \Delta x)$$

Δy means a change (or *increment*) of y. Thus,

$$\Delta y = \mathrm{f}(x + \Delta x) - y = \mathrm{f}(x + \Delta x) - \mathrm{f}(x)$$

EXAMPLE: Find the distance s that a body that was dropped from rest will fall in the 0.1 second after 2 seconds. The applicable equation is

$$s = gt^2/2 = 16.1t^2$$

(See the chapter on kinematics.)

SOLUTION:

$$\Delta s = \mathrm{f}(t + \Delta t) - s = 16.1(2 + 0.1)^2 - 16.1(2)^2 = 6.60\ \mathrm{ft}$$

Derivatives

Let u and v represent any function of x; let b = a constant; let n = a finite constant.

$$\frac{d(x)}{dx} = 1$$

$$\frac{d(b)}{dx} = 0$$

$$d(b + u) = du$$

$$d(bu) = b\,du$$

$$d(u + v) = du + dv$$

$$d(uv) = u\,dv + v\,du$$

$$d\left(\frac{u}{v}\right) = \frac{v\,du - u\,dv}{v^2}$$

$$d(u^n) = nu^{n-1}du$$

$$d\left(\frac{1}{u}\right) = -\frac{du}{u^2}$$

$$d(e^u) = e^u du$$

$$d(\ln u) = \frac{du}{u}$$

$$d \sin u = \cos u\,du$$

$$d \cos u = -\sin u\,du$$

Maxima and Minima

If $y = \mathrm{f}(x)$ is a continuous curve, the function is a maximum where the curve tops out ($dy/dx = 0$) and a minimum where it bottoms out ($dy/dx = 0$). The tangent is a horizontal line at both points. Whether the point is a maximum or minimum in physical problems usually can be logically deduced. In case of doubt, if $x = a$ is the value of x when $dy/dx = 0$,

$$\text{f}''(a) \text{ is negative (concave down) at the maximum}$$

$$\text{f}''(a) \text{ is positive (concave up) at the minimum}$$

If the value of this second derivative is 0, but the third derivative, $\text{f}'''(a)$, does not equal 0, the point at $x = a$ is neither a maximum nor minimum, but is a *point of inflection* where the slope is 0.

Partial Derivatives

A function $z = \text{f}(x, y)$ can be evaluated for given values of x and y. It is often desirable to differentiate a function, holding one of the variables, x or y, constant. This operation is indicated in several ways, such as

$$\left(\frac{dz}{dx}\right)_y = \left(\frac{d\,\text{f}(x,y)}{dx}\right)_y$$

$$\left(\frac{\partial z}{\partial x}\right)_y = \left[\frac{\partial\,\text{f}(x,y)}{\partial x}\right]_y$$

These notations say the function is being differentiated with respect to variable x while y is held constant. The standard notation for partial differentiation is the ∂ sign, but it is not always used.

EXAMPLE: $c_p = \left(\frac{dh}{dT}\right)_p$ or $c_p = \left(\frac{\partial h}{\partial T}\right)_p$ means that the constant pressure specific heat c_p is equal to the change of enthalpy h with respect to the absolute temperature T during a process with the pressure p remaining constant.

INTEGRAL CALCULUS

Integration is the inverse operation of differentiation. Integrations are made indefinite (using the constant of integration) or made between definite limits, as is convenient in solving particular problems.

Indefinite Integrals

The following integrals are frequently used. (a, n, and C are constants.)

A table of integrals is advisable.

$$\int a\,dx = a\int dx = ax + C$$

$$\int x^n dx = \frac{x^{n+1}}{n+1} + C \quad [\text{when } n \neq -1]$$

$$\int \frac{dx}{x} = \log_e x + C \quad \left[x^n dx = \frac{dx}{x} \text{ when } n = -1\right]$$

$$\int \sin x\,dx = -\cos x + C$$

$$\int \cos x\,dx = \sin x + C$$

$$\int e^x dx = e^x + C$$

Areas

To integrate for areas bounded by curves whose equations are known, choose a suitable area dA. If the shaded area in figure 2.8(a) is desired, for example, a horizontal dA would *not* be the first choice because the right end is bounded by two different curves, $x = a$ and $y = f(x)$. The vertical dA shown gives

$$A = \int dA = \int y\,dx = \int_{x_1}^{x_2} \mathrm{f}(x)dx = \int_0^a \mathrm{f}(x)dx$$

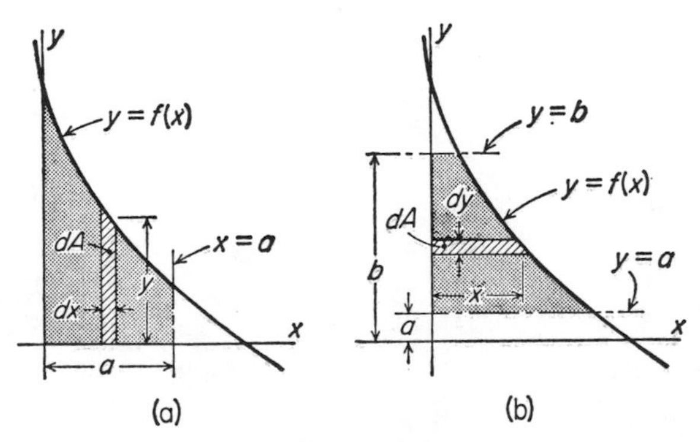

Figure 2.8

For the area in figure 2.8(b), the horizontal dA is more convenient. $x = \mathrm{f}(y)$ must be used because the limits are values of y.

$$A = \int dA = \int x\,dy = \int_{y_1}^{y_2} \mathrm{f}(y)dy = \int_{y_1=a}^{y_2=b} \mathrm{f}(y)dy$$

EXAMPLE: Find the area bounded by the curve $y^2 = 4x$, the y axis, and the lines $y = 2$ and $y = 4$.

SOLUTION: The approach is similar to that shown in figure 2.8(b). From $y^2 = 4x$, we have $x = y^2/4$. Then,

$$A = \int x\,dy = \frac{1}{4}\int_2^4 y^2\,dy = \frac{1}{4}\frac{y^3}{3}\Big]_2^4 = 4.67$$

PRACTICE PROBLEMS

2.1. In a circle of unit radius, sketch a central angle α in the second quadrant, and show lines that are equal in length to $\sin\alpha$, $\cos\alpha$, and $\tan\alpha$.

2.2. Given $\cos 2\alpha = 2\cos^2\alpha - 1$, find $\cos 75°$.

Ans. 0.259

2.3. A man walks 40 feet down the slope of the embankment of a railway which runs due east and west. He then walks 20 feet along the foot of the embankment. He finds that he is exactly NE of the point from which he started at the top of the bank. Express the slope of the embankment in degrees.

Ans. 60°

2.4. $\cos 210°$ equals: (a) $\cos(-30°)$, (b) $\cos 30°$, (c) $-\cos 30°$, (d) $\frac{1}{2}\sqrt{3}$, (e) none of these.

Ans. (c)

2.5. (a) Give the signs of the trigonometric functions (sin, cos, and tan) for each of the four quadrants (upper right, upper left, lower left, and lower right).

(b) Two circles 20 feet in diameter are placed so that the circumference of each just touches the center of the other. Find the area common to both circles.

Ans. 123 ft^2

2.6. The following is given:

$$\cos P - \cos Q = -2\sin\frac{P+Q}{2}\sin\frac{P-Q}{2}$$

Prove that the following is true:

$$\sin(A+B)\sin(A-B) = \cos^2 B - \cos^2 A$$

2.7. $\sec\theta\sin^3\theta\cot\theta$ equals (a) $\cos\theta$, (b) $1+\sin^2\theta$, (c) $1-\cos^2\theta$, (d) $\sin\theta$, (e) none of these.

Ans. (c)

2.8. $1 - \tan^2\theta\cos^2\theta$ equals: (a) $1 - \tan\theta$, (b) $\frac{1}{2}\tan\theta - \sin\theta\cos\theta$, (c) $\cos^2\theta$, (d) $\sin^2\theta$, (e) none of these.

Ans. (c)

2.9. $\log_e e^{2x}$ is equal to: (a) $2x\log_{10} e$, (b) $2\log_e 10$, (c) $2x$, (d) $2 + x$, (e) none of these.

Ans. (c)

2.10. Two tangents to a circle are drawn. They are located so that the distance from each point of tangency to the point of intersection of the two tangents is 20 inches. The chord that connects their points of tangency is 4 inches from the center of the circle. Find the angle between the tangents.

Ans. $50°21'$

2.11. Express $3\sin\omega t + 4\cos\omega t$ in the form $A\sin(\omega t + \alpha)$, giving α in degrees.

Ans. $53°8'$

2.12. A hillside of uniform slope has its greatest inclination, 30°, in the direction due west. A straight roadway in a northerly direction is to be built with a grade of 7%. Find the direction of the road.

Ans. N6°58′W

2.13. Two ships leave the same port at the same time, one sailing due northeast at a rate of 6 mph, and the other sailing due north at the rate of 10 mph. Find the distance between the two ships after they have sailed 3 hours.

Ans. 21.4 mi

2.14. An observer wishes to determine the height of a tower. He takes sights at the top of the tower from points A and B, which are 50 feet apart and on a direct line with the tower. The elevation at point A is 30° and at point B is 40°. What is the height of the tower?

Ans. 92.5 ft

2.15. The difference of two numbers is 7. The difference of their squares is 203. What are the numbers?

Ans. 18, 11

2.16. Given $y = 9$, solve for x in $(x^3)(y^{1.5}) = 8y + 53$.

Ans. 1.67

2.17. At what interest rate will \$80.25 accumulate to \$100.00 in 5 years, interest compounded annually? Solve by logarithms, using the formula

$$S_n = P(1+i)^n$$

S_n is the "amount," or sum of principal and interest at the end of n years, P is the principal invested, and i is the annual interest rate (a decimal).

Ans. 4.5%

2.18. Given that the discharge of water from a pipe is 50 cfs, calculate the number by which this figure should be multiplied to convert it to gpm.

Ans. 449

2.19. Solve for x.

$$2\sqrt{2x+1} = 2x+1$$

Ans. $\frac{3}{2}, -\frac{1}{2}$

2.20. $\log \dfrac{a+b}{c}$ equals:

(a) $\log a + \log b - \log c$, (b) $\frac{\log a + \log b}{\log c}$, (c) $\log(a+b) - \log c$, (d) $e^{(a+b)/c}$, (e) $\log \frac{a}{c} + \log \frac{b}{c}$

Ans. (c)

2.21. Solve for x by logarithms and show each step:

$$x = 5.29(0.896)^{0.435}$$

Ans. 5.0432

2.22. What is meant by a *single valued function*?

2.23. $\log_{10} 360$ equals: (a) 360^{10}, (b) $\log_{10} 60 + \log_{10} 6$, (c) $2\log_{10} 180$, (d) $10\log_{10} 36$, (e) none of these.

Ans. (b)

2.24. Find the value of x in the expression

$$x = \log_e e + 2\log_{10} 100 + \frac{1}{\log_e e^2}$$

Ans. 5.5

2.25. If $R = re^{\tan a}$, then a equals: (a) $\tan^{-1} \log_e \frac{R}{r}$, (b) $\frac{R}{re}$, (c) $\tan^{-1} \frac{R}{re}$, (d) $\log_e \frac{R}{r}$, (e) none of these.

Ans. (a)

2.26. What values of x and y satisfy the following simultaneous equations?

$$2x + 10y = 56$$
$$10x - 6y = 0$$

Ans. 3, 5

2.27. Find the value of the determinant

$$\begin{vmatrix} 5 & 0 & 0 \\ 0 & 1 & 0 \\ 0 & 0 & 3 \end{vmatrix}$$

Ans. 15

2.28. Find the value of $2^2 \times 2^3 + 2$.

Ans. 34

2.29. If 3 is a root of the equation $x^3 - 7x - 6 = 0$, find the remaining roots.

Ans. -1, -2

2.30. Find M if $\log_5 M = \log_{25} 4$.

Ans. 2

2.31. Find the equation of the parabola in the form $y = f(x)$ that passes through the points (1, 2), (3, 20), and (4, 35).

Ans. $y = 2(x + \frac{1}{4})^2 - \frac{9}{8}$

2.32. The acute angle of a right triangle is 35°.

$$\log \sin 35° = 9.75859$$
$$\log \cos 35° = 9.91336$$

The length of the hypotenuse of the triangle is 650 feet and the mantissa of 650 is 0.81291. Find the log of the base, the log of the altitude, and the log tan 35°.

Ans. 2.72627, 2.57150, 9.84523

2.33. To meet each other, A and B started at the same time from two places 180 miles apart. A traveled 6 miles more each day than did B, and the number of miles traveled each day by B was equal to twice the number of days that elapsed before he met A. How many miles did each travel?

Ans. 108 mi, 72 mi

2.34. (a) Find the points of intersection of the following:

$$x - 7y + 25 = 0$$
$$x^2 + y^2 = 25$$

(b) Write the equation of a tangent to the foregoing circle whose slope is $-\frac{3}{4}$.

Ans. (a) $(-4, 3), (3, 4)$, (b) $3x + 4y = 25$

2.35. A lot measures 135.5 feet by 250.5 feet. A walk is to be built around the inside of the lot and is to leave an area of 31,104 sq ft. Find the width of the walk. Write an algebraic equation and show the solution.

Ans. 3.75 ft

2.36. At what points does the line $x + y = 7$ cross the parabola $2y = x^2 + 6$?

Ans. $(-4, 11), (2, 5)$

2.37. Express the equation of the straight line passing through points $x_1 = 2$, $y_1 = 2$ and $x_2 = 4$, $y_2 = 3$ in the form of $y = mx + b$.

Ans. $y = x/2 + 1$

2.38. A straight line is defined by two points. Point A is $x = -2$, $y = -2$. Point B is $x = 2$, $y = 6$. What is the equation of the line?

Ans. $y = 2x + 2$

2.39. What is the approximate length of the line AB in the previous problem?

Ans. $\sqrt{80}$

2.40. A field is a parallelogram with sides 30 and 20 chains long and an area of 42 acres. Find the length of its shorter diagonal. (1 acre = 10 sq ch.)

Ans. 21.05 ch

2.41. Given the equations $x^2 + 4y^2 = 36$ and $16x - x^2 - y^2 = 39$, plot the curves neatly and name them. Solve for the point of intersection, if any.

Ans. (3.62, 2.39), (3.62, −2.39)

2.42. Define *differential* as used in calculus.

2.43. x approaches infinity in the following expression:

$$y = \frac{6x^2 - 2x - 2}{(2x+1)3x}$$

y approaches as a limit: (a) infinity, (b) zero, (c) one, (d) $\frac{2}{3}$, (e) none of these.

Ans. (c)

2.44. The derivative of $(3x^2 + x)^{1/2}$ with respect to x is: (a) $\frac{1}{2}\frac{(6x+1)}{\sqrt{3x^2+x}}$, (b) $\frac{1}{2}(6x+1)$, (c) $\frac{1}{2}\sqrt{3x^2+x}(6x+1)$, (d) none of these.

Ans. (a)

2.45. Given that the acceleration, d^2s/dt^2, of a moving body is a constant, a, develop the formula

$$s = \frac{1}{2}at^2$$

Evaluate the constants of integration from the conditions that $s = 0$ and the body is at rest at $t = 0$.

2.46. Prove that $\int \cos^3 ax\, dx = \frac{1}{a}\sin ax - \frac{1}{3a}\sin^3 ax$.

2.47. t approaches infinity in the following expression:

$$i = \frac{E}{R}(e)^{-t/(RC)}$$

i approaches as a limit; (a) infinity, (b) E/R, (c) one, (d) zero, (e) none of these.

Ans. (d)

2.48. If $A = \int_0^\pi \int_0^R r\, d\phi\, dr$, then A equals: (a) πR^2, (b) πR, (c) $\pi R^2/2$, (d) $\pi R\phi$, (e) none of these.

Ans. (c)

2.49. x, y, and θ are variables in the following parametric equations:

$$x = a(\theta - \sin\theta)$$
$$y = a(1 - \cos\theta)$$

The maximum value of y is: (a) $a(\theta - \sin\theta)$, (b) a, (c) $a - 1$, (d) 0, (e) none of these.

Ans. (e)

2.50. As θ approaches 1 in the expression $x = \frac{1-\theta^2}{1-\theta^3}$, x approaches as a limit: (a) zero, (b) $\tan\theta$, (c) $\frac{2}{3}$, (d) infinity, (e) none of these.

Ans. (c)

2.51. Find the maximum positive value of y on the curve represented by

$$y = 10 - 5(x - 5)^2$$

Ans. 10

2.52. Find the value of $\int_0^{\pi/2} \sin\theta \cos\theta d\theta$.

Ans. $\frac{1}{2}$

2.53. The largest positive value for y on the curve $y = -3x^2 + 6x + 5$ is: (a) 1, (b) 4, (c) 8, (d) 2, (e) none of these.

Ans. (c)

2.54. x approaches zero in the following expression:

$$y = \frac{5x^3 - \cos x + e^x}{x^2 + 2x}$$

What does y approach as a limit?

Ans. $\frac{1}{2}$

2.55. The equation of a curve is $y = 3x^2 + x$. What is the slope of a tangent to this curve at $x = 1$, $y = 4$?

Ans. 7

2.56. Find the area bounded by the curve in the previous problem, the line $x = 2$ and the x axis.

Ans. 10

2.57. The volume of a sphere is increasing at the rate of 6 cu in/sec. Find the rate at which the radius is increasing when the volume is 21 cu in.

Ans. 0.163 ips

2.58. Given a line of fixed length of 10 inches. Find the inclination of this line at which the sum of its horizontal and vertical projections is greatest.

Ans. 45°

2.59. A parabolic bridge arch 8 feet high and 24 feet wide is to be formed as shown in figure 2.9. Determine the length of the inserted brace AB. (*Note*: A is not at mid-span.)

Ans. 7.21 ft

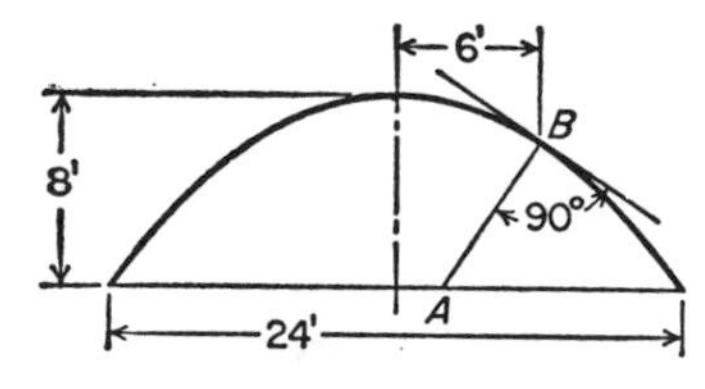

Figure 2.9

2.60. When a plane surface of inclination θ moves through the air horizontally and at a fixed speed, the lifting power of the resisting air is given as

$$Q = K\sin^2\theta\cos\theta$$

Find the value of θ that makes Q a maximum if $K = 1$.

Ans. 54°44′

2.61. A box is to be constructed from a thin sheet of copper 10 inches square by cutting equal squares out of each corner and turning up the remaining parts of the sheet to form the four sides. Use calculus to determine the *largest volume* that can be made from this piece of material by the method described.

Ans. 74 cu in

2.62. The equation of a parabola is

$$x^2 = \frac{b^2}{a}y$$

b is the limit of x and a is the limit of y. Show by calculus that the area B shown in figure 2.10 equals $ab/3$ and that the area C equals $2ab/3$.

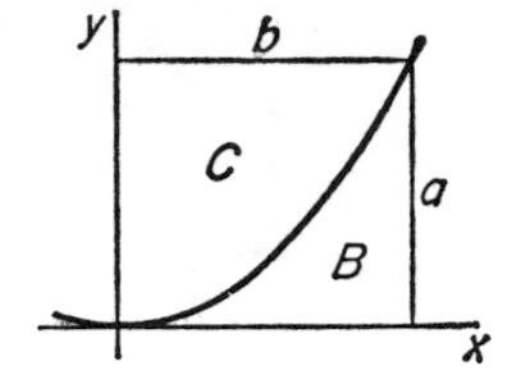

Figure 2.10

3 *CENTROIDS AND MOMENTS OF INERTIA*

SYMBOLS

A = area
g = local acceleration of gravity
g_c = standard gravitational constant (32.2)
I = moment of inertia, rectangular
J = polar moment of inertia
k = radius of gyration
L = length of line or curve
m = mass in slugs ($\text{lbm}/g_c = W/g$)
P_{xy} = product of inertia
r = radius
V = volume
W = weight, force of gravity
$\overline{x}, \overline{y}, \overline{z}$ = coordinates of centroid or center of gravity
ρ = density, lbm/ft^3 or lbm/in^3

Although only a mass has a *center of gravity*, the name is also applied to lines and plane areas, where a better name would be *centroid.*

CENTROID OF LINE

Imagine a line (curve) to have uniform mass per unit of length. The *centroid of the line* is a point with respect to which the imaginary line (with mass) would balance in any position. For a line in a plane, the location of the centroid is defined by

$$\overline{x} = \frac{\int x\, dL}{L}$$

$$\overline{y} = \frac{\int y\, dL}{L}$$

L is the total length of line. This principle can be used, with good approximation, for wires.

CENTROID OF AREA

By Integration

Imagine an area to have a uniform mass per unit area. The *centroid of the area* is a point with respect to which the imaginary area (with mass) would balance in any position. Most engineering problems concern plane areas whose centroid has the coordinates

$$\overline{x} = \frac{\int x\, dA}{A}$$

$$\overline{y} = \frac{\int y\, dA}{A}$$

A is the total area.

EXAMPLE: Locate the centroid of the area bounded by the curve $y^2 = 8x$, the y axis, and the line $y = 8$. See shaded area in figure 3.1.

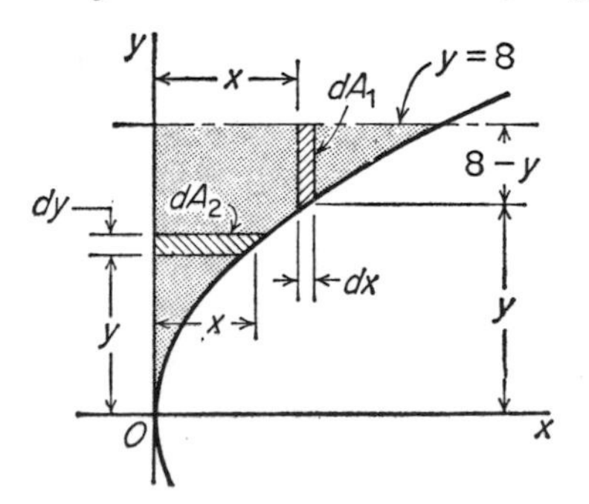

Figure 3.1

SOLUTION: To use the foregoing equations, the total A is needed. It can be obtained from either dA_1 or dA_2 in figure 3.1. Use dA_2.

$$A = \int dA_2 = \int x\,dy = \int_0^8 \frac{y^2}{8}dy = \frac{1}{8}\frac{y^3}{3}\bigg]_0^8 = \frac{64}{3}$$

To get $\overline{y}$, use same dA_2 (moments about the x axis):

$$\int y\,dA_2 = \int_0^8 y\left(\frac{y^2}{8}dy\right) = \frac{1}{8}\frac{y^4}{4}\bigg]_0^8 = 128$$

$$\overline{y} = \frac{\int y\,dA}{A} = \frac{128}{\frac{64}{3}} = 6$$

To get $\overline{x}$, use dA_1. The value of x at $y = 8$ is

$$x = y^2/8 = 8$$

This is the limit of the area in the x direction. Use $y = \sqrt{8}x^{1/2}$. Moments about y axis are:

$$\int x\,dA_1 = \int_0^8 x(8-y)dx = \int_0^8 (8x\,dx - \sqrt{8}x^{3/2}dx)$$

$$= \frac{8x^2}{2} - \frac{\sqrt{8}x^{5/2}}{\frac{5}{2}}\Bigg]_0^8 = 51.2$$

$$\overline{x} = \frac{\int x\,dA}{A} = \frac{51.2}{\frac{64}{3}} = 2.4$$

Composite Areas

If an area is made up of several areas (A'_1, A'_2, etc.), such as triangles and rectangles whose centroids are known,

$$A\overline{x} = A'_1x'_1 + A'_2x'_2 + \ldots = \sum A'x'$$

$$A\overline{y} = A'_1y'_1 + A'_2y'_2 + \ldots = \sum A'y'$$

A'_1 is an area whose known centroid is a distance x'_1 from the axis of moments, etc.

Axes of Symmetry

If there is an *axis of symmetry*, the centroid lies on it. Two axes of symmetry locate the centroid of a plane area. The *centroid of a triangle* lies on the intersection of its medians: one-third of the altitude from a base to the corresponding apex.

EXAMPLE: Determine $\overline{x}$ for the angle section shown in figure 3.2.

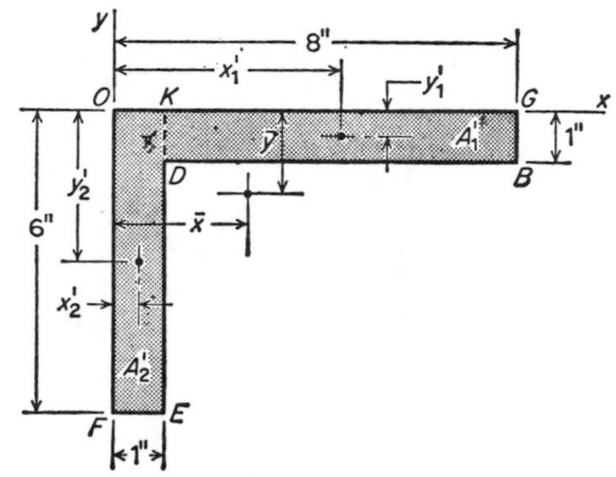

Figure 3.2

SOLUTION: There is no axis of symmetry. Divide into convenient areas (say, rectangles *BDKG* and *KEFO*). Choose any convenient axes, as shown. From the dimensions given, $x_1' = 4.5$ inches, $x_2' = 0.5$ inches, $y_1' = 0.5$ inches, and $y_2' = 3$ inches. Also, $A_1' = 7\text{ in}^2$, $A_2' = 6\text{ in}^2$, and $A = 13\text{ in}^2$.

$$\overline{x} = \frac{A_1'x_1' + A_2'x_2'}{A} = \frac{(7)(4.5) + (6)(0.5)}{13} = 2.65 \text{ inches } (\textit{Ans.})$$

Practice finding $\overline{y}$. (*Ans.* 1.65)

Theorems of Pappus and Guldinus

A *surface* generated by a plane curve (any line) revolving about any non-intersecting axis in its plane has an area of

$$A = \theta\overline{y}L$$

L is the length of generating line, $\overline{y}$ is the distance from the centroid of the *line* to the axis of rotation, θ radians is the angle swept through by any point on the line, and $\theta\overline{y}$ is the distance moved by the centroid. For a complete revolution,

$$\theta = 2\pi$$

If the area is known, this principle can be used to compute $\overline{y}$.

A *volume* generated by a plane area revolving about any nonintersecting axis in its plane has magnitude of

$$V = \theta \overline{y} A$$

A is the magnitude of the revolving area.

EXAMPLE: A torus (doughnut-like shape) is generated by revolving a circular area through 360°. The centroid of both circumference and area is at the center of the circle. Let the radius of this centroid from the axis of rotation be $\overline{y} = R$. r = radius of the revolved circle.

$$\text{surface area of torus}, A = (2\pi R)(2\pi r) = 4\pi^2 Rr$$

$$\text{volume of torus}, V = (2\pi R)(\pi r^2) = 2\pi^2 Rr^2$$

MOMENT OF INERTIA OF AREA

Rectangular Moment of Inertia

Moments of inertia (also called second moments) are taken with respect to planes or axes. There are an infinite number of possible axes for a given area. For a plane area with respect to any rectangular coordinates *in the plane of the area*,

$$I_x = \int y^2 dA \qquad \text{[ABOUT } x \text{ AXIS]}$$

$$I_y = \int x^2 dA \qquad \text{[ABOUT } y \text{ AXIS]}$$

If the axes pass through the centroid, they are the *centroidal moments of inertia*, represented by $\overline{I}_x$ and $\overline{I}_y$.

Moments of inertia are always positive, and have units of linear measurement to the fourth power (e.g., ft^4). Moments of inertia are primarily used in mechanics of materials problems.

Polar Moment of Inertia

The *polar moment of inertia* is the sum of the moments of the elements of the area about an axis perpendicular to the plane of the area. If r is the distance of an element dA from the axis, the polar moment of inertia is

$$J = \int r^2 dA$$

$$\overline{J} = \int r^2 dA$$

The sum of the two rectangular moments of inertia is equal to the polar moment of inertia about the perpendicular axis through the intersection of the rectangular axes.

$$J = I_x + I_y$$

$$\overline{J} = \overline{I}_x + \overline{I}_y$$

Radius of Gyration

Radius of gyration is a mathematical concept that predicts the location of an infinitely-thin line having a moment of inertia equal to that of the actual object.

$$k = \sqrt{\frac{I}{A}}$$

$$k = \sqrt{\frac{J}{A}}$$

Parallel Axis Theorem

This is also called the *transfer formula*:

$$I = \overline{I} + Ad^2$$

$$J = \overline{J} + Ad^2$$

$$k^2 = \overline{k}^2 + d^2$$

Both axes must be parallel.

EXAMPLE: (a) Find the moments of inertia, I_x, I_y, and J_{xy}, for the area of figure 3.1. Let the dimensions be inches.

SOLUTION: For I_y, use dA_1. $y = \sqrt{8x}$.

$$I_y = \int x^2 dA = \int_0^8 x^2 (8 - y) dx$$

$$= \int_0^8 (8x^2 - \sqrt{8}\, x^{5/2}) dx = 8^4/21 = 195.05 \text{ in}^4$$

For I_x, use dA_2 because all its parts are the same distance y from the x axis. $x = y^2/8$.

$$I_x = \int y^2 dA = \int y^2 (x\, dy) = \frac{1}{8} \int_0^8 y^4 dy = 819.2 \text{ in}^4$$

$$J = I_x + I_y = 819.2 + 195.05 = 1014.25 \text{ in}^4$$

(b) What is the polar radius of gyration?

$$k = \sqrt{\frac{J}{A}} = \sqrt{\frac{1014.25}{\frac{64}{3}}} = 6.9 \text{ inches}$$

$A = \frac{64}{3} = 21.33$ from a previous example.

(c) What is the centroidal moment of inertia about an axis parallel to Ox? From the previous example, $\overline{x} = d = 2.4$ inches in the transfer formula:

$$\overline{I}_x = I_x - Ad^2 = 819.2 - (21.33)(2.4)^2 = 696.4 \text{ in}^4$$

Choice of Differential Element

If not too inconvenient, a *differential element* of area parallel to the axis about which the moment of inertia is desired is preferred. An alternative is to choose an element whose parts are not all at the same distance from the axis, but rather an element whose moment of inertia about the axis is known or can be determined easily. You can also choose a dA such as $dx\, dy$ or $\rho\, d\theta\, d\rho$ and make double integrations.

Moments of inertia most frequently useful are:

Circle:

$$\bar{I} = \frac{\pi r^4}{4} = \frac{\pi D^4}{64} = \frac{Ar^2}{4}$$

$$\bar{J} = \frac{\pi r^4}{2} = \frac{\pi D^4}{32} = \frac{Ar^2}{2}$$

Rectangle:

$$\bar{I} = \frac{bh^3}{12} = \frac{Ah^2}{12} \qquad [h = \text{SIDE} \perp \text{AXIS OF } I]$$

Triangle:

$$\bar{I} = \frac{bh^3}{36} \qquad [h = \text{HEIGHT} \perp \text{AXIS OF } I]$$

Product of Inertia

Products of inertia are needed to find principal axes and maximum and minimum values of moments of inertia. Its mathematical definition is

$$P_{xy} = \iint xy\,dA$$

The calculation is a summing of moments about both x and y axes at the same time. If either the x or y axis is an axis of symmetry,

$$P_{xy} = 0$$

Symmetrical or not, x and y are *principal axes* when $P_{xy} = 0$. Given I_x, I_y, and P_{xy} when x and y are *not* principal axes, the angle θ that the principal axes make with the x and y axes is defined by

$$\tan 2\theta = \frac{2P_{xy}}{I_y - I_x}$$

VOLUME

Centroid

The coordinates of the *centroid of a volume* are

$$\overline{x} = \frac{\int x\,dV}{V}$$

$$\overline{y} = \frac{\int y\,dV}{V}$$

$$\overline{z} = \frac{\int z\,dV}{V}$$

In each case, dV is chosen so that all of its parts are the same distance from a coordinate plane; e.g., for $\overline{x}$, all of dV is a distance x from the yz plane. A plane of symmetry contains the centroid. If there are two planes of symmetry, the centroid lies on the line that is the intersection of the planes.

Moment of Inertia of a Volume

$$I = \int r^2 dV$$

r is the distance from dV to the axis of moments. All parts of dV are at the same distance from that axis. See *Choice of Element* for masses.

CENTER OF GRAVITY OF A MASS

The *center of gravity* is the point through which the resultant force of gravity acts. The *center of mass* is located at the same place as the centroid of the volume, provided the material is homogeneous. For engineering purposes, these points are the same. The center of gravity of a material with uniform density is the centroid of volume.

MOMENT OF INERTIA OF A MASS

$$I = \int r^2 dm$$

All parts of dm are at distance r from the axis (or plane) of moments, and the units of I are slug-ft^2.*

The procedure is to establish dV and multiply dV by the density. *Density* is the quantity of matter per unit volume (i.e., the mass per unit volume). In engineering tables, it is usually found in lbm/ft^3 or lbm/in^3, never in slug units. If it is used in lbm/in^3, the volume must be in in^3, etc. The symbol used for mass density is ρ.

Specific weight is the *weight* per unit volume (virtually the same as density when measured close to the earth's surface). Sometimes you see *weight-density*, γ, a confusion of the ideas of weight and density, but you must be prepared for it.

Specific gravity for solids and liquids is the ratio of the mass of a unit volume of the material to the mass of a unit volume of water at a standard temperature (usually 60°F for engineers).

Suppose the density of steel is $\rho = 490$ lbm/ft^3. Then,

$$dm = \frac{\rho}{g_c} dV = \frac{490}{32.2} dV \text{ slugs}$$

$$m = \frac{\rho}{g_c} V \text{ slugs}$$

Radius of Gyration

$$\text{radius of gyration,} \quad k = \sqrt{\frac{I}{m}}$$

$$I = mk^2$$

k is the distance from the axis to a point at which the entire mass can be imagined as concentrated with no change in moment of inertia.

---

*A *slug* is equivalent to lbf-sec^2/ft. A weight in pounds divided by the standard acceleration of gravity ($g = 32.17\ldots$; use 32.2 ordinarily) gives the mass in slugs; say,

$$\frac{w \text{ lbf}}{g \frac{\text{ft}}{\text{sec}^2}} = m \text{ slugs}$$

Choice of Element

The element dm (and dV) used must be one whose

- parts are all at the same distance from the moment axis, or
- moment of inertia about the moment axis is known, or
- moment of inertia about its own centroidal axis that is parallel to the moment axis is known.

The *parallel axis theorem* (*transfer formula*) $[dI = d\bar{I} + (dm)d^2]$ is used and the integration made.

Composite Volumes

For masses, I is always positive. The moment of inertia of a mass with a void is its moment of inertia, considering the void as solid minus the moment of inertia for the void, both *with respect to the same axis.* The transfer formula, $I = \bar{I} + md^2$, applies.

USEFUL VALUES

($m = W/g$ = mass of the body in slugs)

cylinder of radius r about geometric axis:	$\bar{I} = \frac{mr^2}{2}$
hollow cylinder of outside radius r_o and inside radius r_i about geometric axis:	$\bar{I} = \frac{m}{2}(r_o^2 + r_i^2)$
sphere of radius r about a diameter:	$\bar{I} = \frac{2}{5}mr^2$
slender rod of length L about centroidal axis perpendicular to its length:	$\bar{I} = \frac{mL^2}{12}$
thin circular disk of radius r about a "diameter":	$\bar{I} = \frac{mr^2}{4}$
a mass relatively concentrated at some distance d from the moment axis:	$I \approx md^2$

PRACTICE PROBLEMS

3.1. The area shown in figure 3.3 is symmetrical about the vertical axis Y–Y. Compute (a) the distances a and b, locating the horizontal axis X–X, which passes through the center of gravity, (b) the moment of inertia of the area about axis Y–Y, (c) the radius of gyration about axis Y–Y.

Ans. (a) 9.9, 11.1 inches, (b) 1512 in^4, (c) 2.898 inches

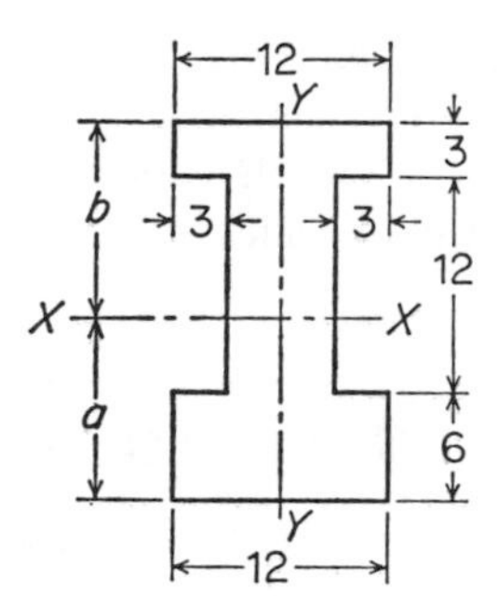

Figure 3.3

3.2. By integration, find the volume of a right circular cone of base radius r and height h.

Ans. $\pi r^2 h/3$

3.3. Calculate the moments of inertia and radii of gyration about both axes of a W14 section with 1 × 15-inch plates welded symmetrically to each flange. Properties of the section are: $I_{xx} = 967$ in^4, $I_{yy} = 350$ in^4, $A = 25.6$ in^2, $d = 14$ inches.

Ans. 2657 in^4, 6.9 inches, 912.5 in^4, 4.05 inches

3.4. Calculate the polar moment of inertia of the cross section of a hollow shaft that is 5 inches OD and 4 inches ID.

Ans. 36.2 in^4

3.5. Find by integration the area generated by revolving around the x

axis the portion of the parabola $y^2 = 9x$ between $x = 1$ and $x = 16$.

Ans. 910

3.6. Find the second moment (moment of inertia) of the area shown in figure 3.4 about its horizontal centroidal axis.

Ans. 90.7 in^4

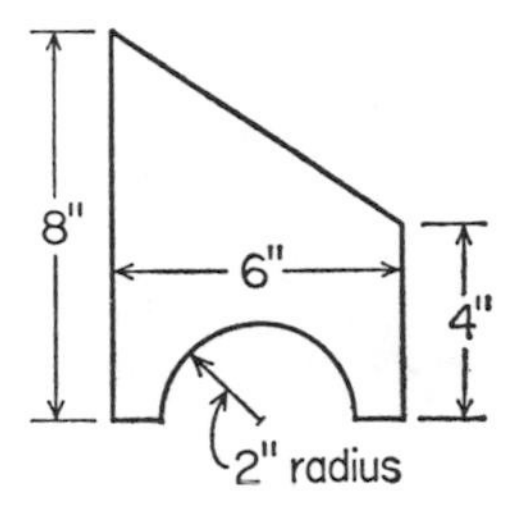

Figure 3.4

3.7. Two homogeneous spheres, A and B, are connected by a rod so that the axis of the rod and the centers of the two spheres are in one horizontal line. Sphere A is 8 inches in diameter and weighs 80 lbm. Sphere B is 4 inches in diameter and weighs 20 lbm. The rod weighs 15 lbm and its diameter can be neglected. The distance from center to center of the spheres is 24 inches. How far from the center of sphere A is the composite center of gravity?

Ans. 5.87 inches

3.8. Show that for a rectangle, $k = d/\sqrt{12}$.

3.9. An I-beam is fabricated from steel with a $\frac{1}{2} \times 6$-inch upper flange, a $\frac{1}{2} \times 4$-inch lower flange, and a $\frac{1}{2} \times 12$-inch web. Find its center of gravity and its moment of inertia about a horizontal line passing through its center of gravity.

Ans. 5.93 inches from top, 264 in^4

3.10. The moment of inertia of an area of 5 in^2 about an axis 2 inches above and parallel to the neutral axis is 100 in^4. What is the moment of inertia of this area about an axis 3 inches above and parallel to the neutral axis?

Ans. 125 in^4

3.11. A steel right circular cone, 9 inches in diameter at the base and 12 inches high, has a concentric hole 4 inches in diameter bored upward from the base. The hole is then filled with lead. How far up must this cylinder be bored into the cone so that, after being filled with lead, the composite center of gravity will be $\frac{1}{16}$ inch closer to the base than the center of gravity for a solid steel cone would be? Steel weighs 0.285 lbm/in^3. Lead weighs 0.41 lbm/in^3.

Ans. 1.43 inches, 4.45 inches

4 *VECTORS*

DEFINITIONS

A *vector* quantity has magnitude, sense, and location. A *scalar* quantity has only magnitude. Mass, volume, and energy are examples of scalars, which are added and subtracted arithmetically. Force, displacement, velocity, acceleration, momentum, and impulse are vectors. A vector not drawn in its true position, as for adding or subtracting, is said to be a *free vector*. In figure 4.1(a), if we say vector $\mathbf{F} = 750$ lbf and $\theta = 30°$, it is defined. In specifying the direction of a vector in writing, we often indicate its sense by the angle from the positive x axis [e.g., ψ for vector $\mathbf{v}_1$ and ϕ for vector $\mathbf{v}_2$ in figure 4.1(b)].

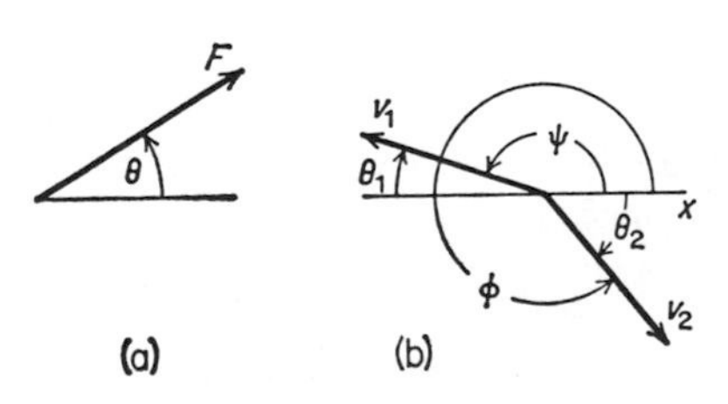

Figure 4.1

Adding Vectors; Components

Suppose there are three vectors, $\mathbf{a}_1$, $\mathbf{a}_2$, and $\mathbf{a}_3$, whose magnitudes and senses are known. Vectors are added by placing tail to point, as in figure 4.2, starting from a convenient pole, O. After the last vector has been added, the vector $\mathbf{R}$ at $\beta°$ and $\theta°$ that goes from O to r is the sum, called the *resultant.*

Vectors that add up to a resultant vector can be said to be the *components* of the resultant. $\mathbf{a}_1$, $\mathbf{a}_2$, and $\mathbf{a}_3$ in figure 4.2 are components of $\mathbf{R}$. In figure 4.3(a), $\mathbf{F}_1$ and $\mathbf{F}_2$ are components of $\mathbf{F}$. The lines of action of two components of a total vector always *intersect on the line of action of the resultant.* Most useful components are rectangular components, at right angles, as in figure 4.3(b), but not necessarily horizontal and vertical.

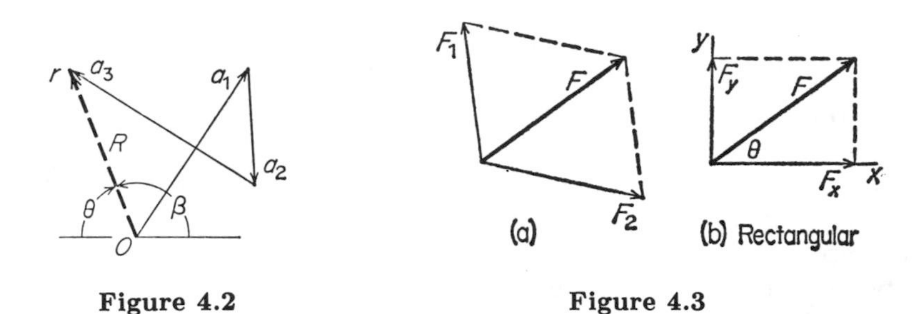

Figure 4.2

Figure 4.3

The process of getting components of a given vector is called *resolution.* $\mathbf{F}$ is resolved into components $\mathbf{F}_x = \mathbf{F}\cos\theta$ and $\mathbf{F}_y = \mathbf{F}\sin\theta$ (figure 4.3(b)). The direction of a vector with respect to its rectangular components is with respect to its x component, as $\theta = \tan^{-1}\mathbf{F}_y/\mathbf{F}_x$ in figure 4.3(b).

Subtracting Vectors

With respect to force analyses, vectors are nearly always summed. With respect to motion concepts, velocity, acceleration, and so forth, they are often subtracted. Subtract vectors by placing them point to point, or tail to tail.

A vector is also subtracted from another by reversing its sense and adding the reversed vector, as seen at BM and QN in figure 4.4.

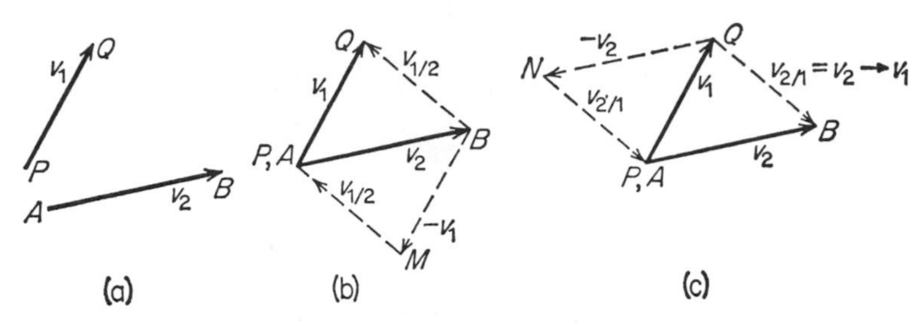

Figure 4.4

RESULTANT, ALGEBRAIC METHOD

Given a system of vectors in a plane, **A**, **B**, **C**, and so forth, as in figure 4.5, the resultant **R** is defined by the following equations:

$$\mathbf{R}_x = \sum X = \mathbf{A}\cos\alpha + \mathbf{B}\cos\beta - \mathbf{C}\cos\gamma$$

$$\mathbf{R}_y = \sum Y = -\mathbf{A}\sin\alpha + \mathbf{B}\sin\beta + \mathbf{C}\sin\gamma$$

$$\theta = \tan^{-1}\frac{\sum X}{\sum Y}$$

$$\mathbf{R}r = \sum M_o = \mathbf{A}a - \mathbf{B}b - \mathbf{C}c$$

$\sum X$ means the sum of the x components of the vectors **A**, **B**, **C**, and so forth; similarly for $\sum Y$. In the last equation, O is any convenient center of moments. From this equation, find the distance r of vector **R** from O. In taking moments of **A**, **B**, **C**, and so forth, all of these vectors must be in their true line of action. If the vectors are parallel, only one sum, in the direction of the vectors, needs to be taken. If the vectors represent forces and $\sum F_x = 0$, $\sum F_y = 0$, but $\sum M_o \neq 0$, then the resultant is a *couple.*

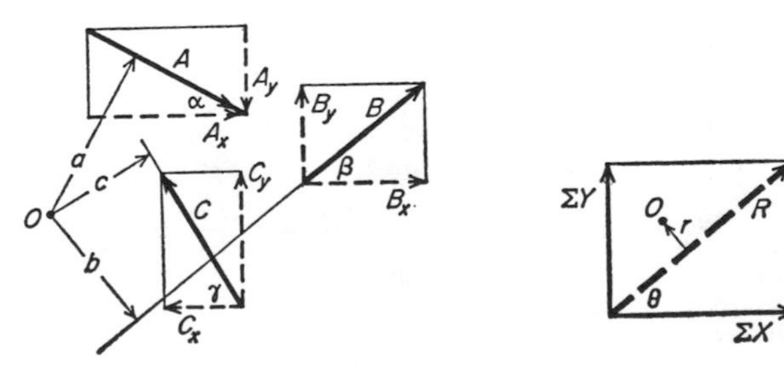

Figure 4.5

5 *KINEMATICS*

SYMBOLS

a, a_x, a_y = acceleration (motion)
a_n = normal acceleration
a_t = tangential acceleration
D = diameter
g = acceleration of gravity
n = revolutions per minute, rpm
r = radius
s, x, y = displacement
t = time
v, v_x, v_y = speed or velocity
α = angular acceleration
θ = angular displacement
ω = angular velocity

Kinematics, a branch of physics, is the science of motion. Motion problems can be solved in terms of scalars (algebraic and calculus approach), or by vector algebra.

Plane Motion

There are three classifications of plane motion for a *point* or *particle*:

rectilinear: Moves in a straight line
rotation: Moves in circular path
curvilinear: Moves in curved path. (Implies other than circular.)

A finite rigid body has:

rectilinear translation: All particles move in straight lines
curvilinear translation: Particles move in curved paths, but body has no angular displacement
rotation: All particles move in circular path (motion about a fixed axis)
general plane motion: Its particles have curvilinear motion. Body's motion can be analyzed as a combination of translation and rotation for infinitesimal movements if not for finite motion.

ALGEBRAIC APPROACH

Motion Concepts

It is helpful to define the motion concepts in terms of differentials, so that only the principles of calculus and other mathematics need to be used.

Velocity is defined by the equations:

$$v = \frac{ds}{dt}$$
$$v_x = \frac{dx}{dt}$$
$$v_y = \frac{dy}{dt}$$

Acceleration is defined by the equations:

$$a = \frac{dv}{dt} = \frac{d^2s}{dt^2}$$
$$a_x = \frac{dv_x}{dt} = \frac{d^2x}{dt^2}$$
$$a_y = \frac{dv_y}{dt} = \frac{d^2y}{dt^2}$$

These equations can be used for any point or particle whose motion (or its x and y components) can be defined to give integrable and differentiable expressions. Another useful form is derived by eliminating dt from the definitions of v and a:

$$v\,dv = a\,ds$$

This is helpful when a is a function of s:

Constant Acceleration

There is no need to memorize the equations for constant acceleration, because they are easy to derive with a = constant.

$$\int_{v_0}^{v} dv = \int_0^t a\,dt = v - v_0 = at$$

$$v = v_0 + at = \frac{ds}{dt}$$

$$\int_0^s ds = \int_0^t v_0\,dt + \int_0^t at\,dt = s = v_0 t + \frac{at^2}{2}$$

The speed is v_0 when $t = 0$, but $s = 0$ when $t = 0$. If a body is falling freely in a vacuum (close to the earth),

$$a = g = 32.2 \text{ ft/sec}^2$$

The derived equations apply. Also, for a constant,

$$\int_{v_0}^{v} v\,dv = \int_0^s a\,ds = \frac{v^2}{2} - \frac{v_0^2}{2} = as$$

For a falling body ($a = g$) for which $v_0 = 0$ when $s = 0$,

$$v^2 = 2gs$$

Analogous equations for component motions in x and y directions are easily derived. Always assume a positive direction.

All foregoing concepts are vectors in the same direction (as horizontal or vertical), but not the same sense. (Some might point up, and some down.) A vector pointing opposite to the chosen positive sense is negative.

Variable Acceleration

EXAMPLE: If $v = 8t + 6t^2$, what are a and s?

SOLUTION:

$$a = \frac{dv}{dt} = 8 + 12t$$

This value can be found for any time t. Since $v = ds/dt$,

$$\int ds = \int (8t\,dt + 6t^2\,dt)$$

$$s = 4t^2 + 2t^3 + C$$

C is a constant of integration which is to be determined from given conditions (any simultaneous values of s and t).

Projectiles

A projectile's motion in a vacuum and near the earth is defined by

$$a_x = 0$$

$$ay = -g = dv_y/dt$$

(Therefore, $v_x = C = v\cos\theta$.) Then, referring to figure 5.1,

$$\frac{dx}{dt} = v\cos\theta$$

$$x = (v\cos\theta)t + C_1$$

$$x = vt\cos\theta$$

$C_1 = 0$ because $x = 0$ when $t = 0$.

$$\int dv_y = -\int g\,dt$$

$$v_y = -gt + C_2 = -gt + v\sin\theta$$

When $t = 0$, $v_y = v\sin\theta$. Therefore,

$$C_2 = v\sin\theta$$

$$\int dy = -\int gt\,dt + v\sin\theta\int dt$$

$$y = -\frac{gt^2}{2} + (v\sin\theta)t + C_3 = -\frac{gt^2}{2} + (v\sin\theta)t$$

When $t = 0$, $y = 0$. Therefore,

$$C_3 = 0$$

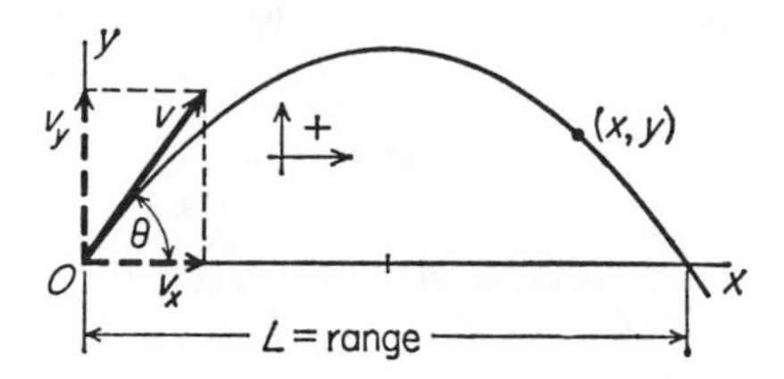

Figure 5.1 Trajectory

Angular Motion

$$\text{angular velocity, } \omega = \frac{d\theta}{dt}$$

$$\text{angular acceleration, } \alpha = \frac{d\omega}{dt} = \frac{d^2\theta}{dt^2}$$

θ is angular displacement (radians). Also, analogous to $v\,dv = a\,ds$,

$$\omega\,d\omega = \alpha\,d\theta$$

Since an arc $s = r\theta$,

$$ds = r\,d\theta$$

r is constant. Hence,

$$v = r\omega$$

$$a = r\alpha$$

A point in a rotating body has a speed of

$$v = 2\pi rn = \pi Dn = r\omega$$

Tangential and Normal Accelerations

The tangential acceleration is the time rate of change of the speed (tangential):

$$a_t = \frac{dv}{dt}$$

Sense accords with α.

The normal acceleration is always directed normal to the path toward the center of the path of the point whose normal acceleration is desired:

$$a_n = \frac{v^2}{r} = r\omega^2 = v\omega$$

The total or resultant acceleration of a point or particle is

$$a = \sqrt{a_t^2 + a_n^2}$$

The total or resultant velocity and acceleration are understood to be with respect to the frame of the machine, which might or might not be attached to the earth.

GRAPHICAL SOLUTIONS

Instantaneous Center

Instantaneous center is also called *instant center, centro, virtual center,* and *rotopole.* If the axis of rotation of a body is known, a graphical construction can be devised that satisfies the rule: *The speeds of points in a rotating body are proportional to their radii.* $v_1/r_1 = v_2/r_2$ for points 1 and 2. The sense of v is always perpendicular to the radius.

Relative Velocity

Velocity problems can be more easily solved by the principle of relative velocities. $v_{A/R}$ is read as the "velocity of point A relative to a reference body R" that is nearly always either the earth or the frame of a moving machine (as a vehicle). $v_{A/B}$ is the velocity of A with respect to B, and so forth. In texts and problem work, the symbol R is generally omitted and the usual reference body is understood. If A and B are two points in a rigid body, the only possible motion of one relative to the other is perpendicular to the line AB between them (i.e., a motion of rotation). Hence,

$$v_{A/B} = (AB)\omega \qquad \text{[}A\text{ AND }B\text{ IN RIGID BODY]}$$

ω is the angular velocity of the body. A vector equation will yield two unknowns: two magnitudes, two directions, or one magnitude and one direction. If points A and B are not fixed in the body, both the magnitude and direction of each must be known to find the magnitude and sense of the relative velocity.

EXAMPLE: A ship is moving directly northeast at 11.52 mph in still water. A man walks across the deck facing due east at 4 mph relative to the ship. What is the man's absolute velocity?

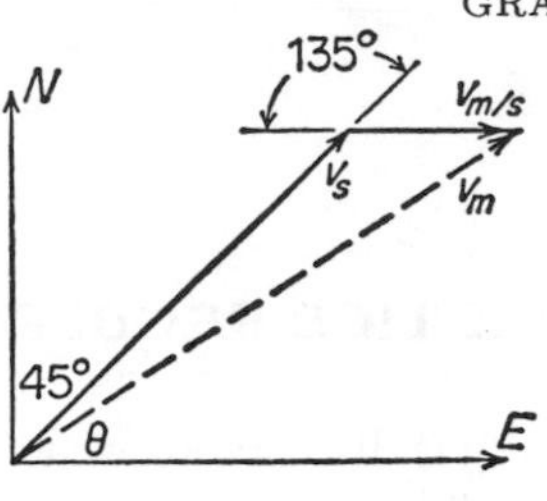

Figure 5.2

SOLUTION: Let v_s = ship's velocity, v_m = man's velocity, and $v_{m/s}$ = velocity of man relative to ship. The graphical solution of the following equation is indicated in figure 5.2:

$$v_m = v_s + v_{m/s} \qquad \text{[VECTOR ADDITION]}$$

The sense is $\theta = 33.9°$ north of east. An algebraic solution can be obtained by using the law of cosines.

Relative Acceleration

$$a_{A/R} = a_{B/R} + a_{A/B}$$

[VECTOR ADDITION; POINTS A AND B FIXED IN RIGID BODY]

Interpretation of subscripts is as explained for velocities. Each acceleration is resolved into tangential and normal components to make this equation useful. Dropping the R,

$$a_{tA} + a_{nA} = a_{tB} + a_{nB} + a_{tA/B} + a_{nA/B}$$

Each tangential acceleration equals $r\alpha$ (which equals $r_A\alpha$ for point A). The relative value is

$$a_{tA/B} = (AB)\alpha$$

Each normal acceleration equals $\frac{v^2}{r}$ (say, $\frac{v^2}{r_A}$ for point A). The relative value is

$$a_{nA/B} = \frac{v^2_{A/B}}{AB} = (AB)\omega^2$$

α and ω define the angular motion of the body that contains points A and B fixed in the body.

PRACTICE PROBLEMS

5.1. A faster-moving body is directly approaching a slower-moving body going in the same direction. The speed of the faster body is V_1, that of the slower body is V_2. At the instant the separation of the two bodies is d, the faster body is given a constant deceleration of a. Determine the formula for the minimum value of d that will prevent a collision.

Ans. $d = -(V_1 - V_2)^2/(2a)$

5.2. A bullet leaves a gun muzzle at a velocity of 2700 ft/sec at an angle of 45° to the horizontal. Determine (a) the maximum height to which the bullet will travel, (b) the maximum distance the bullet will travel horizontally measured along the same elevation as the gun muzzle. Neglect air resistance.

Ans. (a) 56,600 ft, (b) 226,400 ft

5.3. An apparatus for determining the velocity of a rifle bullet consists of two paper disks, mounted 5.0 feet apart on a horizontal shaft which turns at 1750 rpm. A bullet is fired so that it pierces the disks at a radius of 6 inches and the angle measured between the bullet hole in the first disk and the bullet hole in the second disk is 18°0′. What is the average velocity of the bullet during the time the bullet passes from disk 1 to disk 2?

Ans. 2920 ft/sec

5.4. The rifling in a .30-caliber rifle causes the bullet to turn one revolution per 10 inches of barrel length. (a) If the muzzle velocity of the bullet is 2900 ft/sec, how many revolutions per minute is the bullet rotating at the instant it leaves the muzzle? (b) Assuming constant acceleration for the bullet for the 22 inches of barrel travel, what is the magnitude of the acceleration?

Ans. (a) 208,800 rpm, (b) 2.29 EE6 ft/sec^2

5.5. A balloon is 200 feet above the ground and is rising vertically at the constant rate of 15 ft/sec. An automobile passes beneath it traveling along a straight road at a constant rate of 45 mph. How fast is the distance between them changing 1 second later?

Ans. 33.7 ft/sec

5.6. Ship A is traveling due east at a speed of 10 knots. Ship B is 6 nautical miles south and 8 nautical miles west of A and is capable of a speed of 20 knots. What heading should B take to intercept A, and how long will it take?

Ans. N70°48′E, 54 minutes

5.7. A train's speed increases uniformly from 30 mph to 60 mph in 5 minutes while traveling along a straight horizontal track. What is the train's acceleration?

Ans. 0.147 ft/sec^2

5.8. A wheel is rotating about a fixed horizontal axis at its geometric center. The spokes have a radius of r feet. If a spoke turns through $\theta°$, how far does a point on the rim travel?

Ans. $\pi r\theta/180$

5.9. An object thrown upward will return to earth with the magnitude of the terminal velocity equal to (a) the initial velocity, (b) zero, (c) half the initial velocity, (d) none of these.

Ans. (a)

5.10. A submarine 20 miles SW of a harbor sees a ship leave the harbor traveling S80°E at a rate of 12 mph. In which direction and at what rate must the submarine travel in order to meet the ship in 2 hours?

Ans. N75°E, 19.5 mph

5.11. A ball is projected vertically downward from the top of a tower 600 feet high with an initial velocity of 30 ft/sec. At the same instant, another ball is projected vertically upward from the bottom of the tower with an initial velocity of 170 ft/sec. Neglect the effect of wind resistance. Find the distance from the bottom of the tower at which they pass. What is the relative velocity between them?

Ans. 365.1 ft, 200 ft/sec

6 *STATICS*

SYMBOLS

F = force, or frictional force
F_x = x component of F
f = coefficient of friction
M = moment
M_x = moment about x axis
N = normal force
R = resultant force

Statics is a study of systems of forces in equilibrium, in accordance with Newton's first law: Every particle remains (in equilibrium) in a state of rest or moves with a constant velocity (in a straight line) unless an unbalanced force acts on it. If all the external forces on a rigid body add vectorially to zero ($R = 0$), there is no resultant force and the body is in equilibrium. Force systems are:

colinear: All vectors on same line
concurrent: All vectors intersect at a particular point
parallel: All vectors are parallel.

nonconcurrent: Any other

coplanar: All vectors lie in the same plane, and can be any one of the foregoing.

noncoplanar: Force vectors in space. They can be concurrent, parallel, or other.

Conditions of equilibrium for forces in space are:

$$\sum F_x = 0$$
$$\sum F_y = 0$$
$$\sum F_z = 0$$
$$\sum M_x = 0$$
$$\sum M_y = 0$$
$$\sum M_z = 0$$

Certain conditions do not apply to certain force systems.

COPLANAR SYSTEMS

Equilibrium Equations

Independent conditions for equilibrium are:

colinear:

$$\sum F = 0$$

concurrent:

$$\sum F_x = 0$$
$$\sum F_y = 0$$

The x and y directions are any two convenient directions.

parallel:

$$\sum F = 0$$
$$\sum M = 0$$

The moments can be summed about any point in the plane.

non-concurrent:

$$\sum F_x = 0$$
$$\sum F_y = 0$$
$$\sum M_o = 0$$

$$\sum F_x = 0$$
$$\sum M_A = 0$$
$$\sum M_B = 0$$

$$\sum M_o = 0$$
$$\sum M_A = 0$$
$$\sum M_B = 0$$

Subscripts o, A, and B designate three suitable axes about which moments are taken. If you know how to use them, the equations $\sum F = 0$ and $\sum M = 0$ sum up the whole subject of statics.

Free-Body

To use the simple equations of statics successfully, one must be able to make a free-body diagram that shows the actions of other bodies on the (free-) body being studied. Relevant fundamental principals include:

1. The force of gravity always acts through the center of gravity and acts downward. (It is often small enough to be neglected.)

2. The direction of the reaction between smooth (frictionless) surfaces is normal to the surfaces at the point or line of contact.

3. A two-force member can be cut and the action of one part on the other can be replaced by a force (along the member). Two forces in equilibrium are necessarily colinear.

4. If a member with three or more (noncolinear) forces on it is cut, both a moment (couple) and a force generally must be placed at the cut section to maintain equilibrium.

5. A flexible member (e.g., a cord or belt) with forces on it is necessarily a two-force member and is necessarily in tension.

6. The moment arm of a force is always perpendicular to the line of action.

7. The sum of the moments of the components of a force is equal to the moment of the resultant, all about the same axis. (This is *Varignon's theorem.*)

8. The moment of a couple is a specific value. The moment of a force depends on the location of the axis of moments.

EXAMPLE 1: Find the magnitude, direction, and distance from point A of the force required to act on the rigid body in figure 6.1 in order to maintain the body in equilibrium.

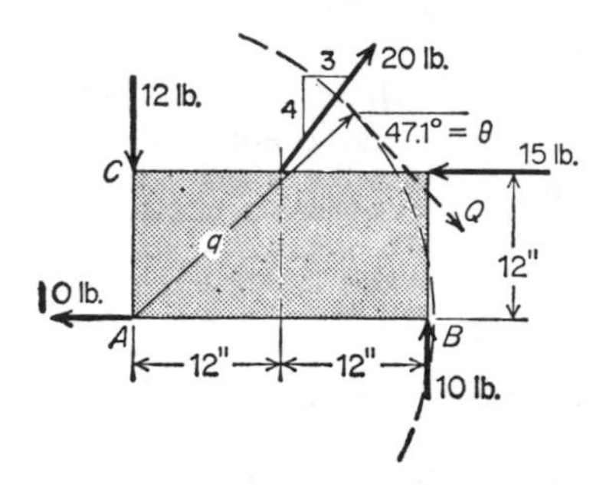

Figure 6.1

SOLUTION: Call the unknown force Q. The magnitude and sense of Q at θ are defined by $\sum F_x = 0$ and $\sum F_y = 0$. Figure 6.1 is the free-body diagram, except that, in the beginning, Q is not shown because its properties are unknown.

$$\sum F_x = Q\cos\theta - 15 - 10 + 20 \times \frac{3}{5} = 0$$

$$\sum F_y = Q\sin\theta - 12 + 20 \times \frac{4}{5} + 10 = 0$$

$$\frac{Q\sin\theta}{Q\cos\theta} = \frac{Q_y}{Q_x} = -\frac{14}{13} = -1.078$$

The acute angle whose tangent is -1.078 is a negative angle (in fourth quadrant) of $\theta = -47.1°$ (*Ans.*)

$$Q = \sqrt{14^2 + 13^2} = 19.1 \text{ lbf} \,|\!\searrow \quad (Ans.)$$

$$\sum M_A = Qq - (10)(24) - (15)(12) + (12)(12) - (16)(12) = 0$$

[CLOCKWISE POSITIVE]

From this,

$$Qq = +468$$

The positive sign shows that the force Q must be located so that it produces a clockwise moment about A at a distance from A of

$$q = 468/19.1 = 24.5 \text{ inches} \quad (Ans.)$$

With the center at A, draw a circle of radius 24.5 inches. Then, draw a tangent at an angle of $-47.1°$ and observe that it must be on the upper side of the circle, as shown, to provide a clockwise moment. (The question does not tell us to show the force on its correct line of action, but it is prudent to complete the solution by doing so.) The resultant of the original force system is $-Q$.

EXAMPLE 2: Given the truss as shown in figure 6.2(a), find the forces in L_oU_2, L_oL_2, U_2U_4, and U_2L_4.

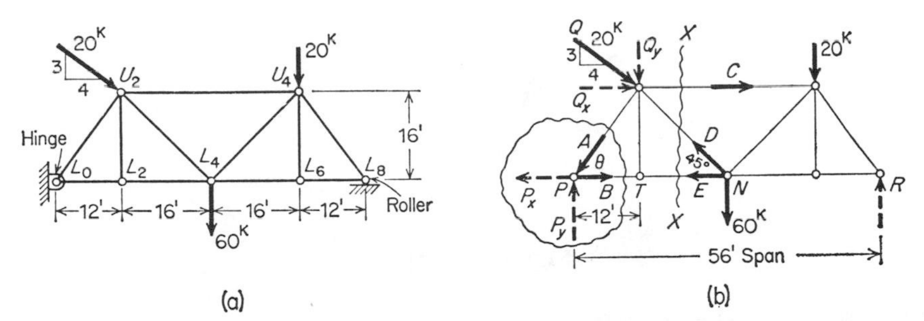

Figure 6.2

Call the desired forces A, B, C, and D to simplify the symbols. First, get support reactions (at least, at the left). It is safer to find all so that a check can be made. Three free-bodies are needed: (1) the whole structure, (2) the pin P, and (3) the section to the left (or right) of X-X. There is only a vertical reaction, R, at the roller. The left reaction is easiest represented by its components, P_x and P_y. $Q_x = 16$ kips and $Q_y = 12$ kips are used instead of Q. For whole structure,

$$\sum M_P = 56R - (44)(20) - (28)(60) - (12)(12) - (16)(16) = 0$$
$$R = 52.8 \text{ kips}$$
$$\sum M_R = 56P_y - (28)(60) - (12)(20) - (44)(12) + (16)(16) = 0$$
$$P_y = 39.2 \text{ kips}$$
$$\sum F_y = 39.2 - 12 - 60 - 20 - 52.8 = 0 \quad \text{(check)}$$
$$\sum F_x = Q_x - P_x = 16 - P_x = 0$$
$$P_x = 16 \text{ kips}$$

Pin P in figure 6.2(b) is a free-body ($\sin\theta = 0.8$, $\cos\theta = 0.6$). It is fairly easy to see that the senses shown for A and B are the correct ones.

$$\sum F_y = P_y - A\sin\theta = 39.2 - 0.8A = 0$$
$$A = 49 \text{ kips, compression} \quad (Ans.)$$
$$\sum F_x = B - A\cos\theta - P_x = 0$$
$$B = 45.4 \text{ kips, tension} \quad (Ans.)$$

Think of pin T as a free-body. Note that the vertical member TQ (theoretically) cannot have a force on it because it cannot be equalized at T.

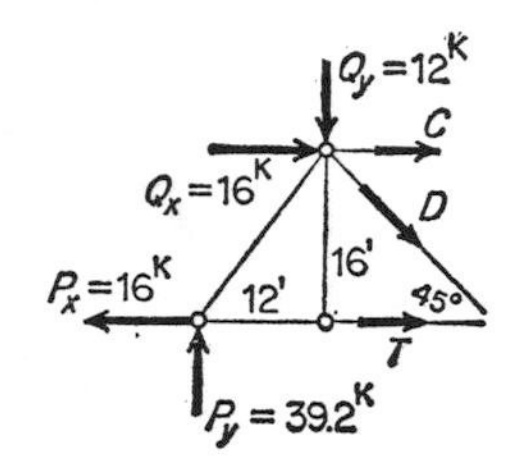

Figure 6.2(c)

The left section of figure 6.2(c) is a free-body. A section must not cut more than three two-force members whose internal forces are unknown. When the sense of an unknown force is not easily decided, put it in as a tensile force. A negative answer shows that it is actually compressive. (Note that A and B are not forces on *this* free-body. Make a separate free-body diagram for each series of equations because this reduces errors.)

$$\sum M_Q = 12P_y + 16P_x - 16T = (12)(39.2) + (16)(16) - 16T = 0$$
$$T = 45.4 \text{ kips, tension}$$

It can be observed directly from the free-body of T that B and T must be the same.

$$\sum F_y = P_y - Q_y - D\sin 45 = 39.2 - 12 - 0.707D = 0$$

$$D = 38.5 \text{ kips, tension} \quad (\textit{Ans.})$$

$$\sum F_x = C + Q_x - P_x + T + D\cos 45 = C + 16 - 16 + 45.4 + 27.2 = 0$$

$$C = -72.6 \text{ kips}$$

$$C = 72.6 \text{ k, compression} \quad (\textit{Ans.})$$

On paper, mark out the arrowhead showing tension for C and insert one for compression, as it is shown on the right of X-X in figure 6.2(b).

The foregoing answers can be checked in several ways. A moment equation about N for the left section would help and should equal zero. Something is wrong if the moment equation does not equal zero. A good check would be to use the right-hand section as a free-body and see if moments, force sums, or both are zero. Do this for practice.

Example 3: The weights A and B in figure 6.3(a) are supported by a continuous rope that is attached at points C and D and that passes around frictionless pulleys as shown. Neglect the weight of the rope and pulleys. Find the position of pulley P relative to point O for equilibrium.

Solution: By implication, the effect of the size of the pulleys is also negligible. Make a sketch and free-body diagram for purposes of a solution, here combined in figure 6.3(b). For a concurrent system in equilibrium at P,

$$\sum F_x = 25\cos\theta - 25\cos\alpha = 0$$

$$\alpha = \theta$$

$$\sum F_y = 25\sin\theta + 25\sin\alpha - 30 = 0$$

$$\sin\theta = \frac{30}{50} = 0.6 = \frac{b}{PD}$$

Triangle PDE is a 3-4-5 triangle.

$$\tan\theta = \tan\alpha = \frac{3}{4} = \frac{b}{a}$$

$$b = \frac{3}{4}a$$

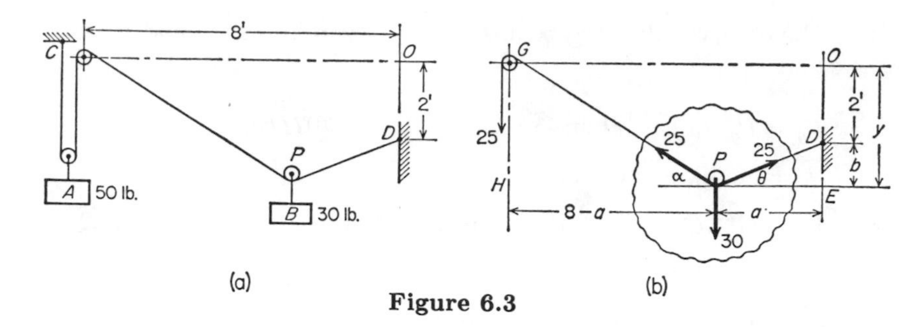

Figure 6.3

From similar triangles PDE and PGH,

$$\frac{b}{a} = \frac{2+b}{8-a}$$

$$\frac{\frac{3}{4}a}{a} = \frac{2+\frac{3}{4}a}{8-a}$$

$a = \frac{8}{3} = 2\frac{2}{3}$ feet and $b = (\frac{3}{4})(\frac{8}{3}) = 2$ feet. The coordinates of P are then $(-2\frac{2}{3}, -4)$ with respect to O. (*Ans.*)

FRICTION

A *smooth* surface is defined as a frictionless one. Actual surfaces are never smooth.

Limiting Friction

If two surfaces in contact have relative motion or tend to have relative motion, a frictional force always acts in a sense to oppose the motion or its tendency to occur. The *coefficient of friction* (most common symbols, f and μ) is defined by

$$f = \frac{F'}{N} = \frac{\textit{limiting}\text{ frictional force}}{\text{normal force}}$$

F' is the *maximum possible* frictional force, but not necessarily the actual frictional force F.

$$F \leq F' = fN$$

If no motion between the contacting surfaces occurs, the frictional force F is just enough to maintain equilibrium and no more. When a body is said to be about to move or on the point of moving, it is assumed that the static limiting frictional force is acting. The coefficient of friction when the contacting surfaces have no relative motion is greater than the coefficient while relative motion is occuring. When relative motion exists, the term *kinetic coefficient of friction*, f_k, is used, and $F = f_k N$ is the frictional force said to be acting. Engineering problems involving friction are uncertain because of the variableness of f and f_k.

Total Plane Reaction

The *limiting angle of friction*, ϕ, is defined by

$$\tan\phi = f(\text{or } f_k) = \frac{F'}{N}$$

The total plane reaction, R, is

$$R = \sqrt{F^2 + N^2}$$

The sense of R is always such that one rectangular component (F) opposes motion, and if motion is impending, its line of action makes an angle ϕ with vector N.

EXAMPLE 1: A block with a weight of 100 lbf rests on a plane inclined 30° with the horizontal. A horizontal force P is applied to the block to push it up the plane. If the coefficient of friction between the plane and the block is $f = 0.3$, what magnitude of P will cause motion to impend up the plane?

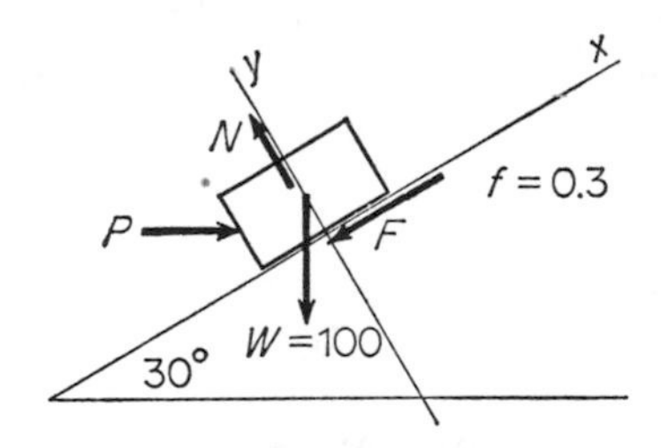

Figure 6.4

SOLUTION: First, make a free-body diagram that also shows the given data, as in figure 6.4. The line of action of P is not specified. Hence, the line of action of N is indeterminate, but the exact locations of these vectors are immaterial to the question in the problem. Reference axes that are parallel and perpendicular to the plane are generally most convenient for inclined-plane problems.

$$\begin{aligned}\sum F_y &= N - W\cos 30^\circ - P\sin 30^\circ \\ &= F/0.3 - 86.6 - 0.5P = 0 \qquad [F = fN] \\ \sum F_x &= P\cos 30^\circ - W\sin 30^\circ - F \\ &= 0.866P - 50 - F = 0\end{aligned}$$

Solve for F from each equation. Equate the results and solve for P.

$$P = 106 \text{ lbf} \quad (\textit{Ans.})$$

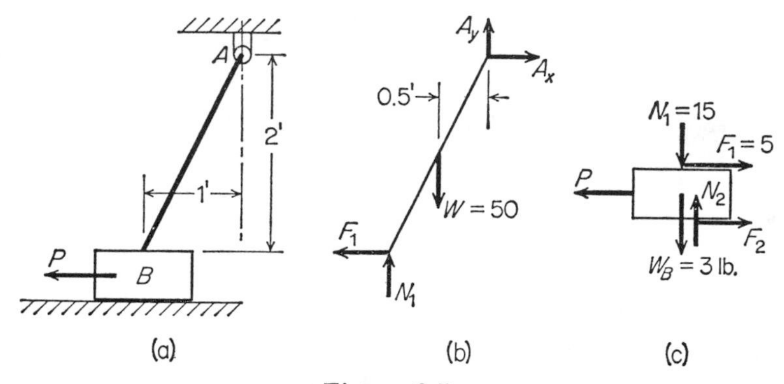

Figure 6.5

EXAMPLE 2: A homogeneous rod weighing 50 lbf is hinged at A in figure 6.5(a) and rests on a block at B. The block, weighing 3 lbf, rests on a horizontal floor. Find the horizontal force P necessary to cause the block to slide to the left if the coefficient of friction for all surfaces is $\frac{1}{3}$. Consider the hinge at A to be frictionless.

SOLUTION: Make a free-body diagram of the rod, as in figure 6.5(b), and take moments about A. Use $F_1 = N_1/3$.

$$\begin{aligned}\sum M_A &= (50)(0.5) - 2F_1 - 1N_1 = 0 \\ \frac{5}{3}N_1 &= 25 \\ N_1 &= 15 \text{ lbf} \\ F_1 &= \frac{15}{3} = 5 \text{ lbf}\end{aligned}$$

Next, make a free-body diagram of the block, as in figure 6.5(c).

$$\sum F_y = N_2 - N_1 - W_B = N_2 - 15 - 3 = 0$$

$$N_2 = 18 \text{ lbf}$$

$$F_2 = fN_2 = \frac{18}{3} = 6 \text{ lbf}$$

$$\sum F_x = F_1 + F_2 - P = 5 + 6 - P = 0$$

$$P = 11 \quad \text{lbf } (Ans.)$$

If static f is exactly $\frac{1}{3}$, P must be slightly greater than 11 lbf to initiate motion.

Belt Friction

If a belt or other flexible member wraps about a drum, pulley, pole, etc., with a contact angle θ (which can be greater than 2π),

$$\frac{T_1}{T_2} = e^{f\theta} \qquad \text{[NO CENTRIFUGAL FORCE]}$$

T_1 is the larger belt tension. T_2 is the smaller force on the other end of the belt. θ is in radians.

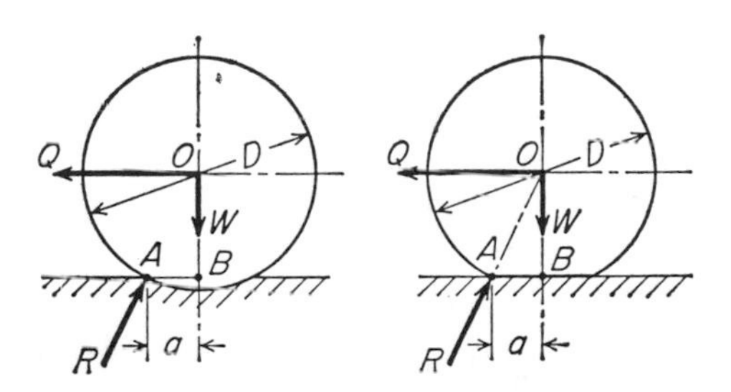

Figure 6.6

Rolling Resistance

Rolling resistance involves no friction but is due to the deformation of the contacting surfaces. The actual situation is a combination of the situations pictured in figure 6.6. (Both members deform.) The ***coefficient of rolling resistance***, a, is as defined by the equation

$$\sum M_A = Wa = \frac{QD}{2}$$

$$Q = \frac{2Wa}{D}$$

Notice, however, that a resistance to rolling is sometimes given to fit the $f = F/N$ concept. Thus, for an equivalent f_r, which might be called the *coefficient of rolling friction* (a misnomer),

$$Q = f_r W$$

The symbols are as in figure 6.6. This coefficient, f_r, must be determined for a particular wheel diameter.

PRACTICE PROBLEMS

6.1. A 260-lbf barrel rests against a rigid vertical wall, A, and is supported by a pin-connected platform that is 30° from the horizontal. The platform is supported by the pin-connected post CD. Dimensions are shown in figure 6.7. Find the reactions at A, B, and D, and the reaction acting on the platform at E.

Ans. 150 lbf, 300 lbf, 173 lbf, and 173 lbf

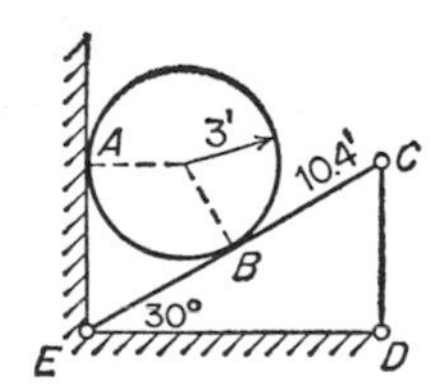

Figure 6.7

6.2. Solve graphically or mathematically for the forces in each member of the truss shown in figure 6.8. Indicate whether the member is in tension or compression.

Ans. AD = 5000 lbf C, BC = 7070 lbf C

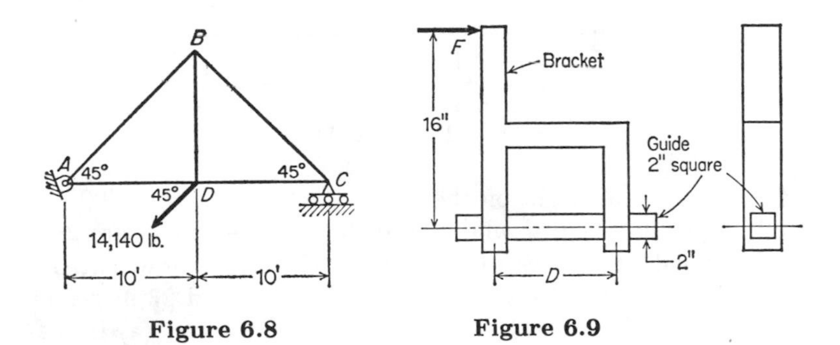

Figure 6.8 **Figure 6.9**

6.3. What is the smallest value of the distance D in figure 6.9 that will cause the bracket to begin to slide along the guide when a force F is applied? Draw a free-body diagram. Use a coefficient of friction of 0.20.

Ans. 6.4 inches

6.4. (a) Draw a free-body diagram of the boom shown in figure 6.10 with the magnitude of all the forces shown. (b) Determine the coefficient of static friction between the boom and wall.

Ans. (b) 0.481

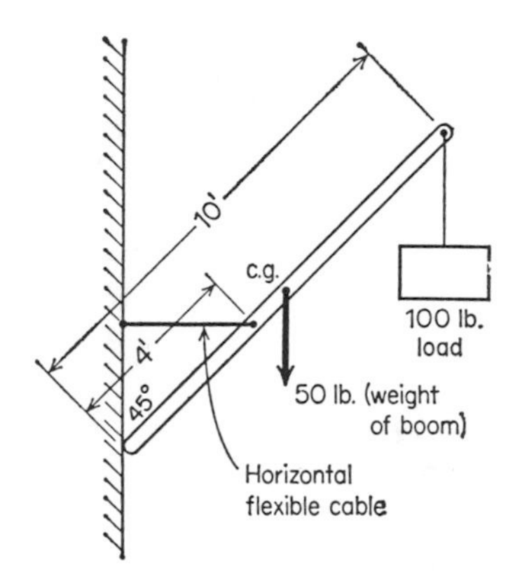

Figure 6.10

6.5. Find the forces in members AB, AC, and BC in figure 6.11 and state whether the member is in tension or compression.

Ans. 4620 lbf C, 6000 lbf T, 4000 lbf C

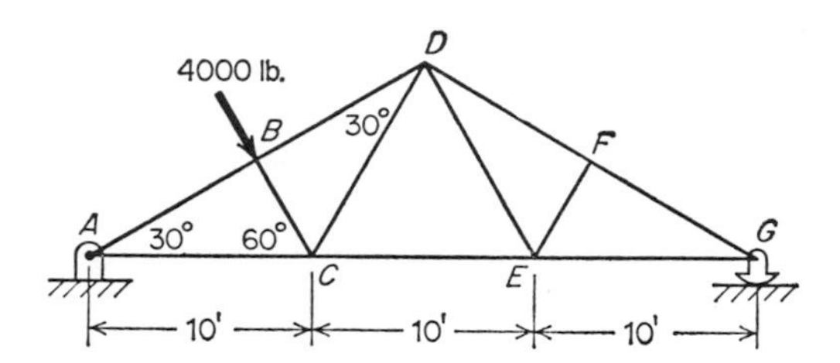

Figure 6.11

6.6. The total weight on the front wheels of a car is 1500 lbf and the total weight on the rear wheels is 1000 lbf. The wheels are 28 inches in diameter and 115 inches from center to center. A draw-bar pull of 750 lbf is exerted horizontally on a trailer hitch located 22 inches above the pavement and 38 inches behind the rear-wheel center. (a) How far is the center of gravity of the car from the front wheels? (b) How much weight must each rear wheel support when the draw-bar pull is 750 lbf on the hitch? (c) What torque must be applied per rear wheel to get 750 lbf draw-bar pull if there is no slipping between the tire and pavement?

Ans. (a) 46 inches, (b) 571.5 lbf, (c) 5250 lbf-in

6.7. A screw jack with 1/8-inch threads is operated by a handle, the grip of which moves through a 9-foot circumference. A force of 20 lbf on the handle will lift what weight, neglecting friction?

Ans. 17,300 lbf

6.8. A system of forces in space is in equilibrium. If two equal and opposite collinear forces are added, which of the following, if any, is true? (a) Equilibrium is destroyed. (b) An unbalance of moment exists. (c) Equilibrium is maintained. (d) None of these is true.

Ans. (c)

6.9. A statement that a given body is in static equilibrium means that the body cannot (a) undergo any displacement, (b) have any acceleration, (c) have any type of motion, (d) be acted upon by more than one force.

Ans. (b)

6.10. A barge to be used in connection with a bridge erection must support two concentrated loads, one of 200 tons and one of 150 tons, located 28 feet apart. The 150-ton load must be 10 feet from one end of the barge. How long must the barge be so that the upward pressure of water is uniform along the full length?

Ans. 52 ft

6.11. The ends of a 21-foot cable are fastened to two rigid overhead points 15 feet apart and in the same horizontal plane. (a) A weight of 200 lbf is attached to the cable 9 feet from one end. Determine the stress (force) in each segment of the cable. (b) If the weight were attached to a roller and permitted to move along the cable, what would be the stress in each segment of the cable?

Ans. (a) 160 lbf, 120 lbf, (b) 143 lbf

6.12. A man weighing 180 lbf stands on a ladder while painting his house. The ladder is 20 feet long and weighs 50 lbf. The man is standing on a rung 4 feet from the top of the ladder. The coefficient of friction between the ladder and wall and between the ladder and ground is 0.25. Determine the maximum angle at which the ladder can be inclined with the vertical before it slips.

Ans. 19.2°

6.13. A pole and anchor are arranged as shown in figure 6.12. Find the force in the guy wire AB and on the pole AC.

Ans. 1275 lbf, 792 lbf

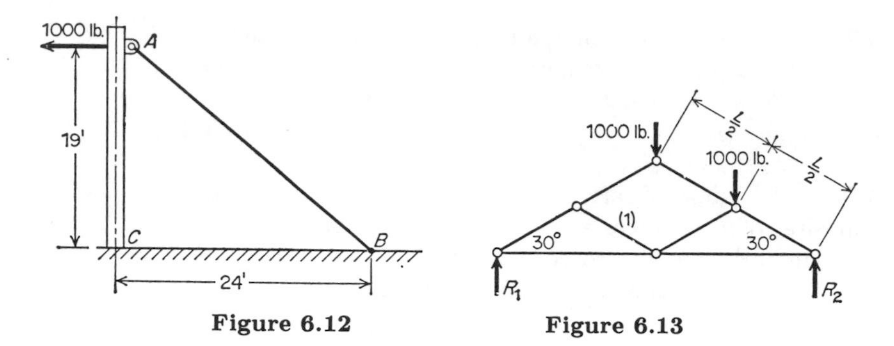

Figure 6.12 **Figure 6.13**

6.14. Find the force on member (1) in figure 6.13 and designate whether it is tension or compression. Compute the values of R_1 and R_2.

Ans. 0 lbf, 750 lbf, 1250 lbf

6.15. A 500-lbf weight is suspended by a rope 50 feet long. A diagonal pull downward at an angle of 45° with the horizontal holds the weight 25 feet from the vertical line through the support at the top. (See figure 6.14.) Determine the tension in the rope, neglecting the rope's weight.

Ans. 1365 lbf

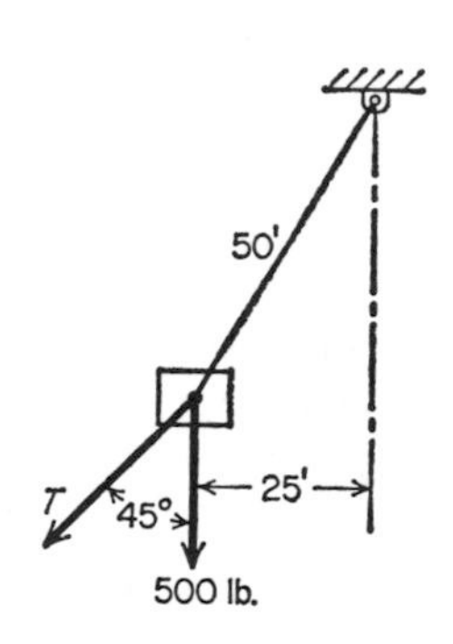

Figure 6.14

6.16. A cylinder 2 feet in diameter and weighing 150 lbf is resting in the 90° angle between the floor and the wall of a box. A force F is applied tangent to the surface, as seen in figure 6.15, directed 45° downward toward the wall. The coefficient of friction between the cylinder and floor and between the cylinder and wall is $f = 0.25$. Find the magnitude of the force F necessary to cause impending motion.

Ans. 66.3 lbf

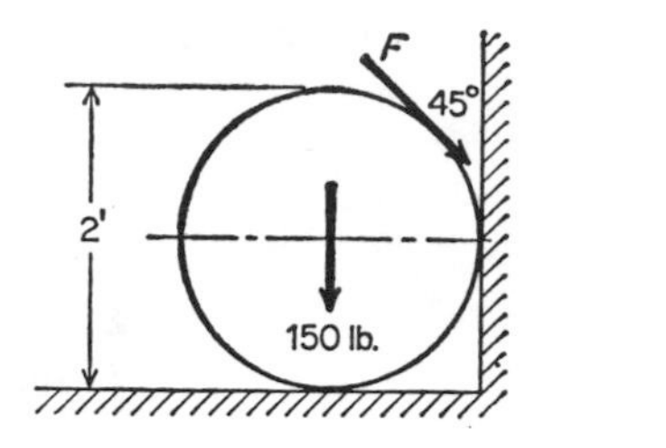

Figure 6.15

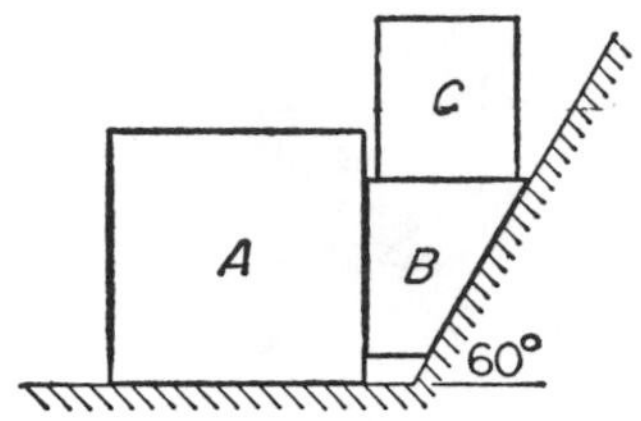

Figure 6.16

6.17. In figure 6.16, C weighs 300 lbf. The coefficient of static friction is $f = 0.3$. Neglect the weight of B. What is the minimum weight of A necessary to prevent motion?

Ans. 740 lbm

6.18. Find the force in member CD in figure 6.17 and state whether it is tension or compression.

Ans. 2598 lbf C

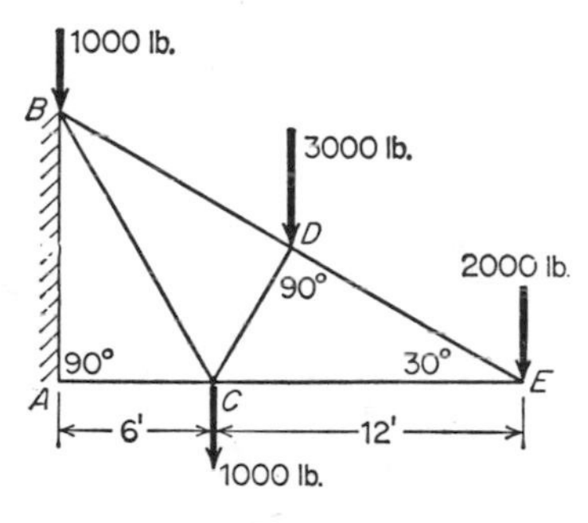

Figure 6.17

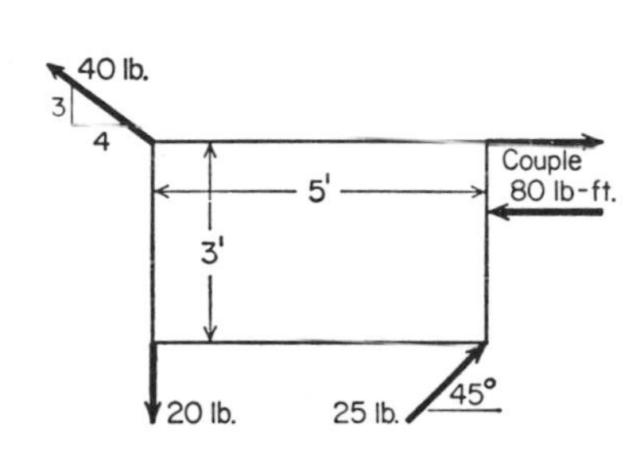

Figure 6.18

6.19. Determine the resultant of the force system shown in figure 6.18.

Ans. 26 lbf

6.20. A horizontal beam 18 feet long is simply supported at the left end and at 6 feet from the right end. It carries a concentrated load of 3000 lbf located midway between the supports. It also carries a load that varies uniformly from zero at the right end to 600 lbf/ft at the left end. Determine the values of the reactions at the two supports. Neglect the weight of the beam. Show all computations and check your results.

Ans. 4200 lbf, 4200 lbf

7 *KINETICS*

SYMBOLS

a = acceleration
$\overline{a}$ = acceleration of center of gravity
d = distance
e = coefficient of restitution
E_k = kinetic energy
E_p = gravitational potential energy of elevation
$\mathbf{F}$ = force, or frictional force
g = local acceleration of gravity
g_c = standard gravitational constant (32.2)
hp = horsepower
i = impulse
I = moment of inertia, slug-ft^2
$\overline{I}$ = moment of inertia about center of gravity axis, slug-ft^2
k = radius of gyration, or spring constant
$\overline{k} = \sqrt{\overline{I}/m}$
M = moment
$\overline{M}$ = moment about center of gravity axis
m = mass in slugs ($\text{lbm}/g_c = W/g$)
$\mathbf{N}$ = normal force
$\mathbf{R}$ = resultant force
r = radius
$\overline{r}$ = radius of center of gravity
s = deformation of a spring
t = time
U = work
v = speed or velocity
$\overline{v}$ = speed or velocity of center of gravity
W = weight in pounds
α = angular acceleration
θ = angular displacement, or angle
ω = angular velocity

The terms *dynamics* and *kinetics* designate the subject matter of this chapter, which deals with force systems that are *not* in equilibrium. There are three techniques for analyzing kinetics problems: (1) force, mass, and acceleration, (2) work and kinetic energy principle, and (3) impulse-momentum principle. Each of these principles is advantageous in certain situations.

Newton's laws can be stated as follows.

1. Every particle remains in a state of rest or moves with constant velocity (in a straight line) unless an unbalanced force acts on it.

2. The acceleration of a particle is directly proportional to the resultant force acting on it and inversely proportional to its mass, and the sense of the acceleration is the same as that of the resultant force.

3. To every action, there is an equal and opposite reaction.

Newton's *law of universal gravitation* states that the attractive force $\mathbf{F}$ between two *particles* is directly proportional to the product of their masses and inversely proportional to the square of the distance d between them:

$$\mathbf{F} = G\frac{m_1 m_2}{d^2}$$

G is the gravitational constant.

$$G = 6.67 \text{ EE-8 dyne-cm}^2/\text{gm}^2$$

For $\mathbf{F}$ lbf, m slugs, and d feet,

$$G = 3.44 \text{ EE-8 ft}^4/\text{lbf-sec}^4$$

FORCE, MASS, AND ACCELERATION

For a body in plane motion, Newton's second law is extended to apply to finite bodies in the following form:

$$\sum \mathbf{F}_x = m\overline{a}_x$$

$$\sum \mathbf{F}_y = m\overline{a}_y$$

$$\sum \overline{M} = \overline{I}\alpha$$

$\sum \mathbf{F}_x$ and $\sum \mathbf{F}_y$ are x and y components of the resultant. In kinetics, the resultant is often called the *effective force.* These equations, properly interpreted, give an infallible approach to the solution of any problem in *plane* motion. However, the center of gravity of a body is not always a convenient center of moments. Hence, as desired, substitute for $\sum \overline{M} = \overline{I}\alpha$ the equation

$$\sum M_o = I_o \alpha$$

The axis of moments O_o is (1) through the center of gravity (as above), (2) the fixed axis of rotation of a rotating body, (3) through a point whose total acceleration is directed through the center of gravity, or (4) through a point whose acceleration is zero.

If a body accelerates in *translation,* $\sum M_o = 0$ when O is the center of gravity or is on the line of action of the resultant.

Convenient x and y axes for a point in a *rotating body* are often the normal and tangential axis.

$$a_t = r\alpha$$

$$a_n = v^2/r = r\omega^2$$

Simulated Equilibrium

This is known as *d'Alembert's principle.* It sometimes is called *dynamic equilibrium* (a contradiction of terms). With $m\overline{a}$ transposed to the left-hand side,

$$\sum \mathbf{F} - m\overline{a} = 0$$

$-m\overline{a}$ is called the *inertia force*, the *dynamic reaction*, or the *reversed effective force* (although it is not really a force). The radial dynamic reaction in a rotating member is called the *centrifugal force.* Similarly, in $\sum \overline{M} - \overline{I}\alpha = 0$, the term $-\overline{I}\alpha$ is the *reversed effective moment*, a dynamic reaction for any body that has an angular acceleration α.

If these dynamic reactions are included in the free-body (i.e., if a vector $-m\overline{a}$ through the center of gravity in the opposite sense to that of $\overline{a}$, and a couple $-\overline{I}\alpha$ in the opposite sense to α, are added to the free-body), the body has been placed in equilibrium. You can then *sum forces in any direction and equate to zero*, or *sum moments about any axis and equate to zero.* This freedom of summing is a frequent convenience. In application, use the components of $-m\bar{a}$ ($-m\bar{a}_x$ and $-m\bar{a}_y$) both passing through the center of gravity. For a body in translation, $-I\alpha$ is zero. Since there are pitfalls in applying any of the equations of this chapter, practice is essential.

Usually it is easier to take the direction of motion as the positive sense, after which *the signs of all vector quantities must be consistent* (**F**, a, M, and α here). The relation between a weight (force of gravity) of W lbf and a mass in lbm is

$$W/g = \frac{\text{lbm}}{g_c}$$

This is true whether the gravitational pull (g) is produced by the earth or another heavenly body.

EXAMPLE 1: A carriage that rolls on a track, AB, carries a heavy load and is represented diagrammatically in figure 7.1. The total weight of the carriage and load is $W = 1000$ lbf. The center of gravity is 8 feet below the track on the vertical center line. A horizontal force $P = 150$ lbf is applied as shown. The resisting forces, **F**, at the wheels are each 0.05 times the corresponding normal forces (assumed to include bearing and rolling resistances). Find the acceleration and the normal and resisting forces.

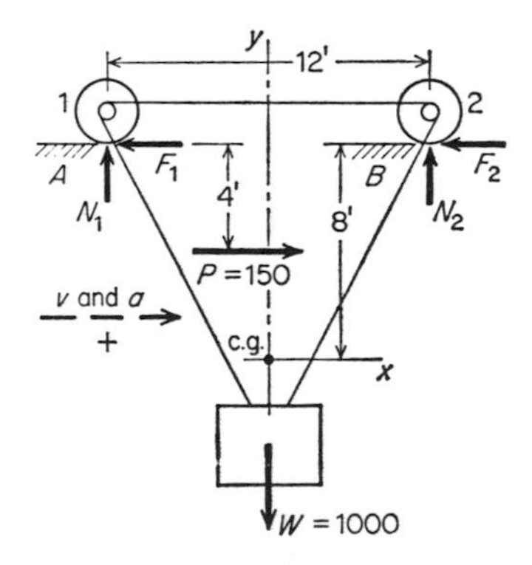

Figure 7.1

FIRST SOLUTION: This solution is without inertia forces on the free-body.

Summing forces ($\overline{a}_y = 0$) and using $\mathbf{F} = 0.05\mathbf{N}$,

$$\sum \mathbf{F}_y = \mathbf{N}_1 + \mathbf{N}_2 - 1000 = 0$$

$$\sum \mathbf{F}_x = P - \mathbf{F}_1 - \mathbf{F}_2 = m\overline{a}$$

$$\mathbf{F}_1 = 0.05\mathbf{N}_1$$

$$\mathbf{F}_2 = 0.05\mathbf{N}_2$$

$$\sum \overline{M} = 6\mathbf{N}_1 - 6\mathbf{N}_2 - 8\mathbf{F}_1 - 8\mathbf{F}_2 + (4)(150) = 0$$

$m = W/g$ slugs. From the equations, the five unknowns (a, $\mathbf{N}_1$, $\mathbf{F}_1$, $\mathbf{N}_2$, and $\mathbf{F}_2$) are found to be

$$\mathbf{N}_1 = 483 \text{ lbf}$$

$$\mathbf{F}_1 = 24.15 \text{ lbf}$$

$$\mathbf{N}_2 = 517 \text{ lbf}$$

$$\mathbf{F}_2 = 25.85 \text{ lbf}$$

$$\overline{a} = 3.22 \text{ ft/sec}^2$$

SECOND SOLUTION: In this solution, the inertia force $-m\overline{a} = 1000\overline{a}/g$ is added to the free-body in figure 7.2 to simulate equilibrium. ($\overline{a}_y = 0$ and $\alpha = 0$, so there is no $-m\overline{a}_y$ or $-\overline{I}\alpha$.) Now, the sum of the forces in any direction and the sum of the moments about *any* point equal zero.

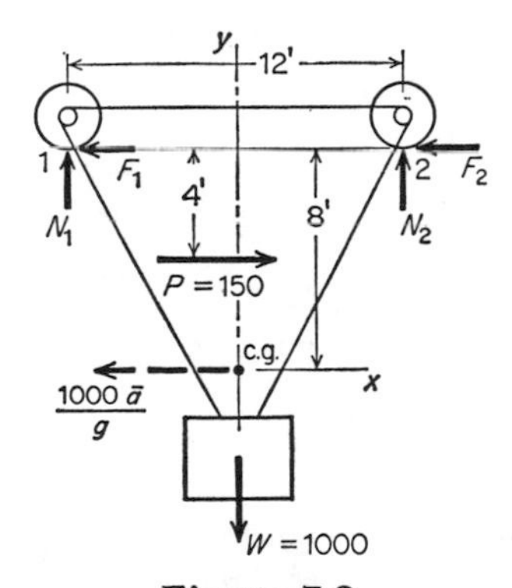

Figure 7.2

(a) $$\sum \mathbf{F}_y = \mathbf{N}_1 + \mathbf{N}_2 - 1000 = 0$$

(b)
$$\sum \mathbf{F}_x = 150 - \mathbf{F}_1 - \mathbf{F}_2 - \frac{1000\overline{a}}{g} = 0$$
$$\mathbf{F}_1 = 0.05\mathbf{N}_1$$
$$\mathbf{F}_2 = 0.05\mathbf{N}_2$$

All are virtually the same as before. However, taking moments about the contact point 1, we eliminate $\mathbf{F}_1$ and $\mathbf{N}_1$.

(c)
$$\sum M_1 = (4)(150) + 12\mathbf{N}_2 - (6)(1000) - (8)\frac{(1000\overline{a})}{32.2} = 0$$

It is also helpful to note that

$$\mathbf{F}_1 + \mathbf{F}_2 = 0.05(\mathbf{N}_1 + \mathbf{N}_2) = 0.05(1000) = 50 \text{ lbf}$$

Equation (a) was used. This value of $\mathbf{F}_1 + \mathbf{F}_2$ into (b) gives $\overline{a} = 3.22$ ft/sec^2. This value of $\overline{a}$ in (c) gives $\mathbf{N}_2 = 517$ lbf. Complete the details of both solutions to see the advantage of the second solution in this particular kind of problem.

EXAMPLE 2: (a) An automobile that weighs 3700 lbf approaches a curve on a concrete-paved highway. The radius of the curve is 1000 feet, and the pavement is flat, with no superelevation. If the coefficient of friction between the tires and the pavement, which is slick, is 0.15, what is the maximum speed the automobile can attain on the curve without skidding sideways? (b) What must the transverse slope of the pavement be to drive at 60 mph on the curve and keep the resultant forces perpendicular to the pavement? Assume the pavement is 24 feet wide and the distance between treads is 6 feet.

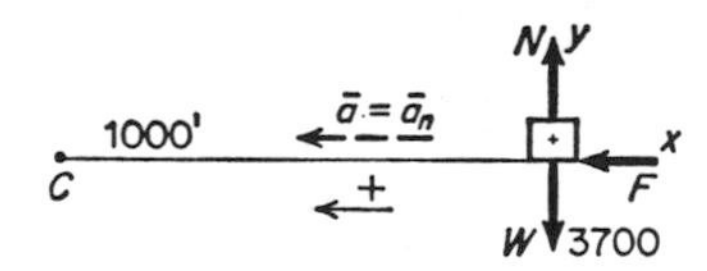

Figure 7.3

SOLUTION: (a) Treat the car as a particle, as in figure 7.3. The center of curvature of the road is C. The only acceleration is the normal acceleration, $\overline{a}_n = v^2/r = v^2/1000$ ft/sec^2.

$$\sum \mathbf{F}_y = \mathbf{N} - 3700 = 0$$
$$\mathbf{N} = 3700 \text{ lbf}$$

$$\mathbf{F} = f\mathbf{N} = (0.15)(3700) = 555 \text{ lbf}$$

$$\sum \mathbf{F}_x = \mathbf{F} = m\bar{a}_n \qquad \text{[LEFTWARD +]}$$

$$= 555 = \frac{3700}{32.2}\frac{v^2}{1000}$$

From this,

$$v = 69.5 \text{ ft/sec}$$

With the inertia force on the free-body and simulated equilibrium, the solution is nearly identical to the above.

(b) This part of the question is somewhat ambiguous. The "resultant force" is always in the sense of the resultant acceleration, which, for constant tangential speed, is always in a radial direction for the car going around the curve. What is meant is that there will be no tendency to slide ($\mathbf{F} = 0$), as in figure 7.4. The total plane reaction is $\mathbf{N}$, perpendicular to the pavement. With this interpretation, the free body in simulated equilibrium has three forces on it: W, $\mathbf{N}$, and ma_n.

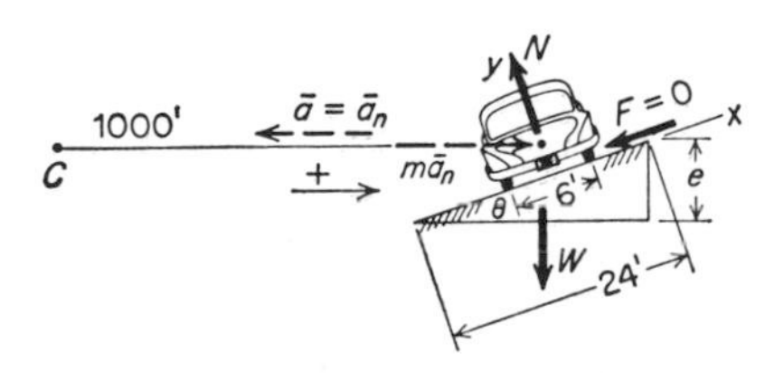

Figure 7.4

$$\sum \mathbf{F}_x = ma_n \cos\theta - W\sin\theta = \frac{Wv^2}{gr}\cos\theta - W\sin\theta = 0$$

From this,

$$\tan\theta = \frac{v^2}{gr} = \frac{88^2}{(32.2)(1000)} = 0.24 \quad (\textit{Ans.})$$

60 mph is 88 ft/sec. In case the angle of slope was intended, you could give $\theta = 13.5°$. The width of pavement is not pertinent to the question (although it would be if the *superelevation*, e, were required), nor is the tread distance pertinent (although it would be if individual wheel reactions were required for the case of finite friction).

Analogous situations include the following: (a) For the conical pendulum in figure 7.5, show that $\tan\theta = v^2/gr$. (b) An airplane is banking on a turn, as in figure 7.6. If there is no side slip, $\mathbf{F}_1 = \mathbf{F}_2 = 0$. If it is slipping outward, some $\mathbf{F}_1$ acts (slight air friction). If it is slipping inward, some $\mathbf{F}_2$ acts (no $\mathbf{F}_1$).

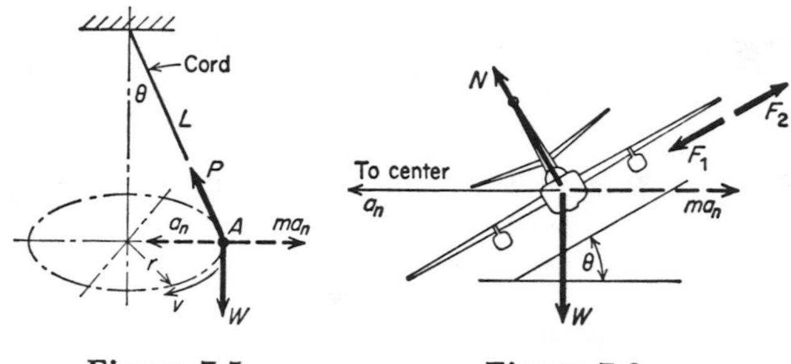

Figure 7.5 **Figure 7.6**

For the automobile on a banked curve, you might show that for $\phi = \tan^{-1} f$:

For impending skidding outward,

$$v = \sqrt{gr \tan(\theta + \phi)}$$

For impending skidding inward (steep bank, slow speed),

$$v = \sqrt{gr \tan(\theta - \phi)}$$

EXAMPLE 3: A 4-foot flywheel, which weighs 3220 lbm and has a radius of gyration of 15 inches, is subjected to a torque of 2400 lbf-in. Determine the tangential acceleration of a point on its rim.

SOLUTION: Taking moments about its center (and center of gravity),

$$\sum \overline{M} = \overline{I}\alpha = m\overline{k}_2\alpha$$

$$\frac{2400}{12} = \left(\frac{3220}{32.2}\right)\left(\frac{15}{12}\right)^2 \alpha$$

$\alpha = 1.28$ radians/sec. At a radius of 2 feet,

$$a_t = r\alpha = (2)(1.28) = 2.56 \text{ ft/sec}^2 \quad (\textit{Ans.})$$

EXAMPLE 4: The 2-foot cylinder in figure 7.7, weighing 322 lbf, is rolled up a rough plane inclined at an angle of 30° with the horizontal by a pull, $P = 184$ lbf, acting through its axis. Compute the acceleration of the center of gravity of the cylinder and the minimum coefficient of friction required for rolling. Rolling resistance is negligible.

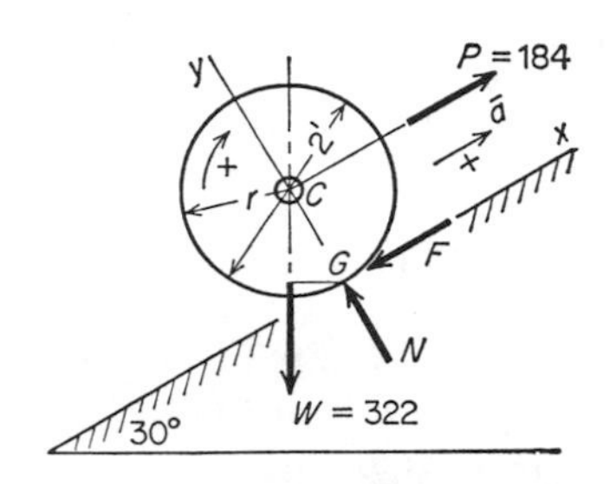

Figure 7.7

SOLUTION: The cylinder is going to move up the plane. In order to prevent relative motion (sliding up) at the point of contact G, the frictional force must act down. In accordance with the rules for center of moments O, take moments either about the center of gravity or the point G of contact with the ground. (Total acceleration of G for a *rolling* wheel is through the center of gravity.) If C in figure 7.7 is used, there will be two unknowns. (Try this anyway.) Since there is one unknown in the moment equation about G, this way is somewhat easier.

$$\sum M_G = I_G \alpha$$

$$I_G = \overline{I} + md^2 = \frac{mr^2}{2} + mr^2 = \frac{3mr^2}{2}$$

$$\sum M_G = Pr - W(r \sin 30^\circ) = \frac{3}{2} \times \frac{Wr^2\alpha}{g} = \frac{3}{2} \times \frac{Wr^2}{g} \times \frac{\overline{a}}{r}$$

$$= 184 - (322)(0.5) = \frac{3}{2} \times \frac{322}{32.2} \times \overline{a}$$

From this, $\overline{a} = 1.533\ \text{ft/sec}^2$. (*Ans.*)

$$\sum \mathbf{F}_x = P - \mathbf{F} - W \sin 30^\circ = \frac{W}{g}\overline{a}$$

$$= 184 - \mathbf{F} - 161 = \frac{322}{32.2} \times (1.533)$$

From this, $\mathbf{F} = 7.67$ lbf. This is the actual frictional force required to prevent sliding.

$$\sum \mathbf{F}_y = \mathbf{N} - W \cos 30^\circ = 0$$

$$\mathbf{N} = 278.5 \text{ lbf}$$

$$f = \frac{\mathbf{F}}{\mathbf{N}} = \frac{7.67}{278.5} = 0.0275 \quad (Ans.)$$

WORK AND KINETIC ENERGY

Energy is a scalar quantity, but it might have algebraic signs, depending on whether the stored energy is increasing or decreasing, or on whether energy is flowing into or out of the system under consideration.

Work

The work, U, done by a force is the product of *the displacement of the point of application of the force times the component of the force in the direction of the displacement*—sometimes shortened to *force times distance.* If θ is the angle of the line of action of the force with the direction of the displacement, as in figure 7.8, the work is

$$U = \int_0^s \mathbf{F} \cos\theta \, ds$$

The component $\mathbf{F}\sin\theta$ does no work; it has no displacement. If both $\mathbf{F}$ and θ are constant,

$$U = (\mathbf{F}\cos\theta)s$$

If several forces (and couples) acting on a body do work, the net work, U_{net}, is the algebraic sum of the works done by the individual forces (and couples), or the work done by the resultant of the system of forces, provided the points of application of all forces in the system undergo the same displacement.

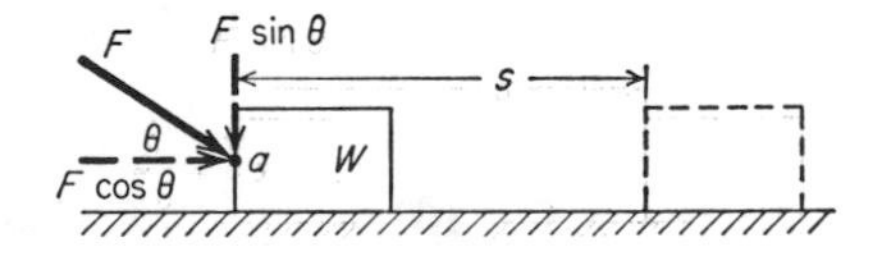

Figure 7.8

The work of a constant couple M is

$$U = M\theta$$

θ is the *total* angular displacement of the couple over a time period.

The work done by or against the force of gravity is often called the change of *potential energy*, ΔE_p, and is *always equal to the force of gravity times the vertical displacement of the center of gravity of the body* or of a system of bodies. In obtaining the net work done, the work of the force of gravity, W, must always be included.

Kinetic Energy

The kinetic energy, E_k, of a body of mass m slugs is the energy arising from the fact that the body is in motion (v ft/sec):

translation: $$E_k = \frac{mv^2}{2} = \frac{Wv^2}{2g} \text{ ft-lbf}$$

rotation: $$E_k = \frac{I_o\omega^2}{2} \text{ ft-lbf}$$

plane motion: $$E_k = \frac{m\overline{v}^2}{2} + \frac{\overline{I}\omega^2}{2} \text{ ft-lbf}$$

$$E_k = \frac{I_o\omega^2}{2} \text{ ft-lbf} \quad [O = \text{ROTATION AXIS}]$$

O is the instant center of rotation, ω is the angular velocity of the body whose kinetic energy is desired, I is in slug-ft^2, and ω is in radians/sec.

Work and Kinetic Energy Principle

This is a special case of the law of conservation of energy, but is derived from Newton's law,

$$\mathbf{F} = ma = mv\,dv/ds$$

If the only forms of energy involved are work and kinetic energy, then the principle is

$$U_{\text{net}} = \Delta E_k$$

The net work done on a rigid body is equal to the change of kinetic energy of the body.

translation: $$U_{\text{net}} = \frac{mv_2^2}{2} - \frac{mv_1^2}{2} = \frac{W}{2g}(v_2^2 - v_1^2) \text{ ft-lbf}$$

rotation: $$U_{\text{net}} = \frac{I_o\omega_2^2}{2} - \frac{I_o\omega_1^2}{2} = \frac{I_o}{2}(\omega_2^2 - \omega_1^2) \text{ ft-lbf}$$

plane motion: $$U_{\text{net}} = E_{k2} - E_{k1} \text{ ft-lbf}$$

See above for E_k for plane motion.

Springs

Springs exert a varying force, a simple variation. Hooke's law for elastic bodies is

$$\mathbf{F} = ky$$

k is the *spring constant* in lbf/inch (for work in lbf-inch) or lbf/ft (for ft-lbf). k is also called the *scale*, the *rate*, or the *modulus*. For a force collinear with the axis of the spring ($\theta = 0$ in $\mathbf{F}\cos\theta$), the work done with a constant k is

$$U = k\int_0^s y\,dy = \frac{ks^2}{2}$$

s is the total deformation, either a compression or an extension, measured from the *free length* (spring without load). The work done by or on the spring for a deformation from s_1 to s_2 is

$$U = k\int_{s1}^{s2} y\,dy = \frac{k}{2}(s_2^2 - s_1^2)$$

As before, $U_{\text{net}} = \Delta E_k$. For springs 1, 2, 3, and so forth, in series and in parallel, the equivalent spring scale, k, is given by

series: $$\frac{1}{k} = \frac{1}{k_1} + \frac{1}{k_2} + \frac{1}{k_3} + \ldots$$

parallel: $$k = k_1 + k_2 + k_3 + \ldots$$

Power

Power is the time rate of doing work. The time must always be indicated (e.g., ft-lbf/minute, BTU/hr). Larger units of power used by engineers are horsepower (hp) and kilowatt (kw).

Handy conversion constants include:

$$33{,}000\ \frac{\text{ft-lbf}}{\text{hp-minute}}$$

$$550\ \frac{\text{ft-lbf}}{\text{hp-sec}}$$

$$0.746\ \frac{\text{kw}}{\text{hp}}$$

$$2544\ \frac{\text{BTU}}{\text{hp-hr}}$$

$$1\ \text{EE7}\ \frac{\text{dyne-cm}}{\text{watt-sec}}$$

$$3412\ \frac{\text{BTU}}{\text{kw-hr}}$$

If the point of application of force **F** moves at the rate of v_m ft/minute or v_s ft/sec in the direction of **F**, power is $\mathbf{F}v_m$ ft-lbf/minute or $\mathbf{F}v_s$ ft-lbf/sec (also said to be the rate of doing work) and

$$hp = \frac{\mathbf{F}v_m}{33{,}000} = \frac{\mathbf{F}v_s}{550} = \frac{M\omega_m}{33{,}000} = \frac{M\omega_s}{550}$$

M = the couple or torque in lbf-ft units, ω = angular velocity, ω_m is radians/minute, and ω_s is radians/sec. Also note that $\omega = 2\pi n$, where n is in revolutions per unit of time, can be used in the foregoing equation.

$$hp = 2\pi nM/33{,}000$$

Efficiency

The *efficiency of a machine* is generally the ratio of output/input. For *thermal efficiency*, see elsewhere in this book.

EXAMPLE 1: A drop hammer of one ton dead weight capacity is propelled downward by a 12-inch diameter cylinder. (a) At 100 psi air pressure, what is the impact velocity if the stroke is 28 inches? (b) What is the impact energy?

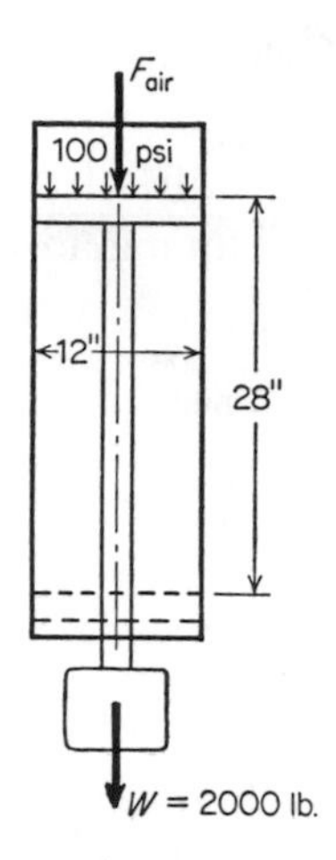

Figure 7.9

SOLUTION: (a) First, sketch the set-up, including the free-body diagram, as in figure 7.9. Because the statement does not say otherwise or give pertinent data, it is logical to assume that the work of frictional forces is negligible, that the air pressure remains constant for the entire stroke, and that the actual movement of the hammer is 28 inches.

$$\text{force of air} = pA = p\frac{\pi D^2}{4} = 100 \times \frac{\pi 144}{4} = 11{,}300 \text{ lbf}$$

$$\text{work of air} = \mathbf{F}s = 11{,}300 \left(\frac{28}{12}\right) = 26{,}400 \text{ ft-lbf}$$

$$\text{work of gravity} = Ws = 2000 \left(\frac{28}{12}\right) = 4670 \text{ ft-lbf}$$

$$\text{net work}, U_{\text{net}} = 31{,}070 \text{ ft-lbf}$$

For $v_1 = 0$ and $U_{\text{net}} = \Delta E_k$,

$$31{,}070 = \frac{mv^2}{2} = \frac{2000}{2g}v^2$$

From this, $v = 31.6$ ft/sec.

(b) Kinetic energy at impact is the same as

$$U_{\text{net}} = E_k = 31{,}070 \text{ ft-lbf}$$

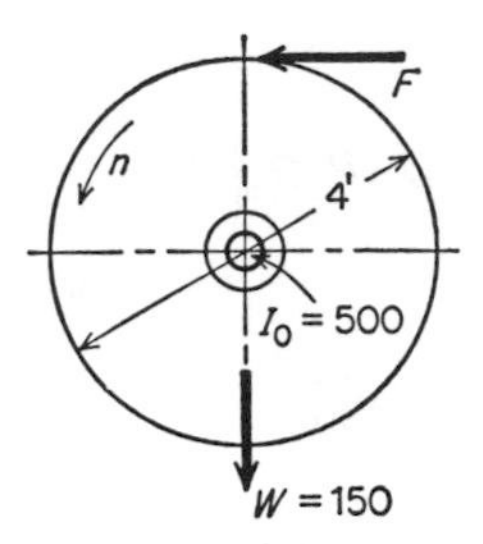

Figure 7.10

EXAMPLE 2: A wheel, 4 feet in diameter and weighing 150 lbf, is mounted on a fixed shaft and has imparted to it from rest a speed of 300 rpm in 1 minute by a constant horizontal force **F** applied tangent to the rim. The axis of the wheel is horizontal and the moment of inertia about the axis is 500 slug-ft^2. Assuming no friction at the shaft, find (a) the magnitude of force **F**, (b) the force of the bearings on the shaft, (c) the kinetic energy at the end of 1 minute. (d) Assuming that when the speed is 300 rpm, force **F** ceases to act and a brake is applied to the shaft, whose moment is 200 in-lbf, find the number of revolutions the wheel makes before stopping.

SOLUTION: First sketch a free-body diagram with appropriate data, as in figure 7.10.

Solutions can be found for either $M = I\alpha$ or $U = \Delta E_k$. Choose work: $\theta = 300 \times 2\pi/2 = 300\pi$ radians during 1 minute, since it increases uniformly. (α is constant if F is.)

$$\begin{aligned} M &= \mathbf{F}r = 2\mathbf{F} \\ \omega_1 &= 0 \\ \omega_2 &= 10\pi \text{ radians/sec} \\ U &= M\theta = (2\mathbf{F})(300\pi) \\ E_k &= \frac{I\omega^2}{2} = \frac{500}{2g} \times 100\pi^2 \end{aligned}$$

Also,

$$600\pi\mathbf{F} = \frac{25{,}000\pi^2}{g}$$

From this,

$$\mathbf{F} = 4.06 \text{ lbf} \quad (Ans.)$$

(b) With no friction, as stated, the force on the bearing is

$$R = \sqrt{\mathbf{F}^2 + W^2} = \sqrt{4.06^2 + 150^2} \approx 150 \text{ lbf} \quad (Ans.)$$

The direction of R is

$$\beta \approx 270^\circ$$

(c) From part (a),

$$E_k = \frac{I\omega^2}{2} = \frac{25{,}000\pi^2}{g} = 7660 \text{ ft-lbf} \quad (Ans.)$$

(d) This part calls for a new free-body diagram, which you should sketch similarly to figure 7.10, except that $\mathbf{F} = 0$ and a couple, $M = 200$ in-lbf clockwise (CW), is added. Then, for counterclockwise as positive,

$$M = I\alpha = -\frac{200}{12} = \frac{500}{g}\alpha$$

$$\alpha = -1.073 \text{ radians/sec CW}$$

(The direction is CW because its measure is negative.) Use $\omega\, d\omega = \alpha\, d\theta$. For $\omega_2 = 0$ and $\omega_1 = 10\pi$ radians/sec,

$$\frac{-\omega_1^2}{2} = \alpha\theta$$

$$\theta = \frac{-\omega_1^2}{2\alpha} = \frac{-100\pi^2}{-1.073 \times 2} = 460 \text{ radians}$$

$$460/(2\pi) = 73.2 \text{ rev.} \quad (Ans.)$$

This part can be solved most readily by work and kinetic energy.

$$M\theta = \Delta E_k$$

EXAMPLE 3: A 4-foot cylinder A has a cord that wraps about a central groove 2 feet in diameter. This cord passes parallel to the 30° incline on which the cylinder rests, over a smooth post E (or frictionless pulley), and then vertically downward. On the end of the vertical part hangs a weight B of 300 lbf. The weight of the cylinder is 500 lbf. Its radius of gyration is $\overline{k} = 1$ foot. The frictional force on the cylinder is sufficient to ensure rolling. (a) If the system is released, in which direction does motion occur?

(b) Determine the velocity of B after it has moved 15 feet from rest.

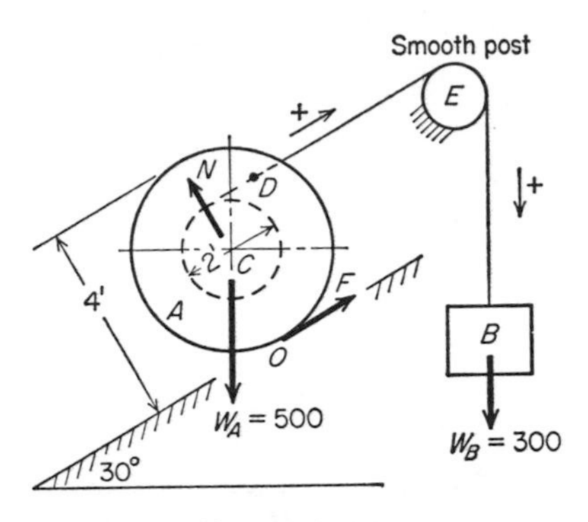

Figure 7.11

SOLUTION: First, convert the verbal description to a sketch, as in figure 7.11, wherein the entire system is the free-body. The instant center of A is point O. Therefore, a point D on the cord moves farther than the center of gravity at C by the ratio

$$\frac{3}{2} = \frac{s_D}{s_c} = \frac{v_D}{v_C} = \frac{a_D}{a_C} = \frac{v_B}{\overline{v}_A} = \frac{a_B}{\overline{a}_A}$$

There are several ways to decide the direction of motion. One is to note that the only forces in figure 7.11 that do work are W_A and W_B, and then to compute the work of each for some finite displacement of the system. Let B move $s_D = s_B = 1$ foot. Then,

$$s_C = 2s_B/3 = \frac{2}{3} \text{ ft}$$

$$U \text{ of } W_B = 300 \times 1 = 300 \text{ ft-lbf}$$

$$U \text{ of } W_A = (500 \sin 30°)\left(\frac{2}{3}\right) = 167 \text{ ft-lbf}$$

Therefore, B predominates and moves down. Take the direction of motion as positive.

(b) In using $U_{\text{net}} = \Delta E_k$, determine the work of each force separately (because they "move through different distances") and sum the changes of kinetic energies.

$$s_C = \frac{2s_B}{3} = 10 \text{ ft}$$

$$\bar{I} = m\bar{k}^2 = \left(\frac{500}{g}\right)1^2 = \frac{500}{g}$$

$$\omega_A = \frac{v_C}{r} = \frac{2v_B}{3r}$$

$$\omega_A^2 = \frac{4v_B^2}{9r^2}$$

$$W_B(15) - (W_A \sin 30°)(10) = \frac{W_B v_B^2}{2g} + \frac{W_A \bar{v}_A^2}{2g} + \frac{\bar{I}\omega_A^2}{2}$$

$$4500 - 2500 = \frac{300}{2g}v_B^2 + \frac{500}{2g}\left(\frac{2}{3}v_B\right)^2 + \frac{500}{2g}\frac{4v_B^2}{9r^2}$$

Therefore, for $r = 2$ feet,

$$v_B = 14.94 \text{ ft/sec}$$

Note that since there is no relative motion at O, the frictional force does no work (when the body is rolling).

The force in the cord might be required. Let it be called Q and imagine the cord cut just above B in figure 7.11. Q is inserted to make a free-body of B. Then, the net work on B is

$$U_{\text{net}} = (300 - Q)15 = \frac{W_B v_B^2}{2g} = \frac{(300)(14.94)^2}{2g}$$

From this,

$$Q = 230.6 \text{ lbf}$$

Without friction at E, the force is the same in all parts of the cord. With friction, it would reduce to Q_2 on the A side of E in accordance with the belt friction equation,

$$Q_1/Q_2 = e^{f\theta}$$

EXAMPLE 4: A 90-lbf body, A, rests on a 30° inclined plane where $f = 0.2$. It is in contact with a spring that has beeen compressed 10 inches and whose scale is $k = 30$ lbf/in. The lower end of the spring is attached to a fixed wall. The body A is released, and when the spring reaches its free length, it ceases to act on A. What is the speed of A at the instant the spring reaches its free length?

SOLUTION: Construct the free-body diagram as shown in figure 7.12.

$$\sum \mathbf{F}_y = \mathbf{N} - W\cos 30° = 0$$
$$\mathbf{N} = (90)(0.866) = 77.94 \text{ lbf}$$
$$\mathbf{F} = f\mathbf{N} = (0.2)(77.94) = 15.59 \text{ lbf}$$

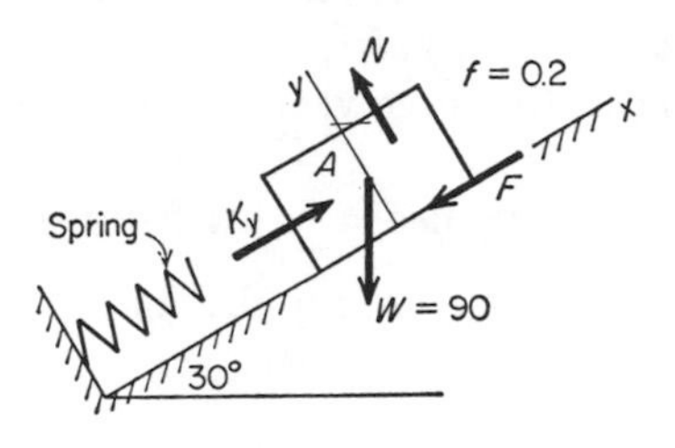

Figure 7.12

Summing the works of the various forces, letting the spring work be positive, for $k = (30)(12) = 360$ lbf/ft,

spring:

$$U_1 = k\int_0^{10/12} y\,dy = \left.\frac{ky^2}{2}\right]_0^{10/12} = \frac{(360)\left(\frac{10}{12}\right)^2}{2} = 125 \text{ ft-lbf}$$

friction:

$$U_2 = -\mathbf{F}s = -15.59\left(\frac{10}{12}\right) = -13 \text{ ft-lbf}$$

gravity:

$$U_3 = -W(s\sin 30°) = -(90)\left(\frac{10}{12}\right)\left(\frac{1}{2}\right) = -37.5 \text{ ft-lbf}$$

$$U_{\text{net}} = 74.5 = \frac{mv^2}{2} = \frac{90v^2}{2g_0}$$

From this,

$$v = 7.3 \quad \text{ft/sec} \quad (Ans.)$$

IMPULSE AND MOMENTUM

Both impulse and momentum are vector quantities.

Impulse

The *linear impulse* i of a force vector $\mathbf{R}$ is the vector $i = \int \mathbf{R}\,dt$. If the force is constant,

$$i = \mathbf{R}\,\Delta t \text{ lbf-sec}$$

$\mathbf{R}$ is a force acting during a time Δt. The moment of the vector of linear impulse is called the *angular impulse*, which is another vector. This is the same as the moment of the force vector times the duration Δt, or, for constant M,

$$\text{angular impulse} = M\,\Delta t \text{ ft-lbf-sec}$$

The frictional force does produce an impulse and must not be omitted.

Momentum

The instantaneous value of *linear momentum* is $m\overline{v}$ lbf-sec, where $\overline{v}$ is the velocity of the center of gravity. *Angular momentum* is the moment of the linear momentum vector about some point.

$$\text{angular momentum} = I_o\omega \text{ ft-lbf-sec}$$

The axis of moments O for this simplified form must be either (1) the instant center of rotation, (2) the mass center, or (3) a "point" whose velocity is directed through the mass center.

Principle of Impulse and Momentum

Using $\mathbf{R} = m\overline{a} = m\,d\overline{v}/dt$ for constant mass, we get

$$\mathbf{R}\,\Delta t = m\overline{v}_2 - m\overline{v}_1 = m\,\Delta\overline{v} \text{ lbf-sec}$$

This means that the impulse is equal to the change of momentum of all vectors. Thus, its use algebraically might be in the forms

$$\mathbf{R}_x\Delta t = m(\overline{v}_{x2} - \overline{v}_{x1})$$
$$\mathbf{R}_y\Delta_t = m(\overline{v}_{y2} - \overline{v}_{y1})$$

m slugs is the mass that undergoes a change of momentum.

$$m = \frac{W}{g} = \frac{\text{lbm}}{g_c}$$

Similarly, for angular impulse and momentum,

$$M_o \Delta t = I_o(\omega_2 - \omega_1)$$

The axis O is as defined above.

This principle also defines force as the time rate of change of momentum:

$$\mathbf{R} = \Delta m\overline{v}/\Delta t$$

This is the way force is created in a jet engine.

Most of the problems in this chapter can be solved by this principle, but impulse-momentum has advantages for certain fluid-flow problems and for impact problems (finite and molecular).

EXAMPLE 1: Gases enter a jet engine at 1500 ft/sec, and exhaust at 3000 ft/sec. What rate of flow is necessary to produce a thrust of 10,000 lbf?

SOLUTION: Let $\Delta t = 1$ sec.

$$\mathbf{F} = \frac{W}{g}(v_2 - v_1) = 10{,}000 = \frac{W}{g}(3000 - 1500)$$

From this,

$$W = 214.5 \text{ lbf/sec} \quad (Ans.)$$

The horsepower being developed is

$$(10{,}000)(1500)/550 = 27{,}300 \text{ hp}$$

EXAMPLE 2: A solid sphere rolls, without slipping, down a 30° plane, starting from rest. After 2 seconds, what is the speed of the center of the sphere?

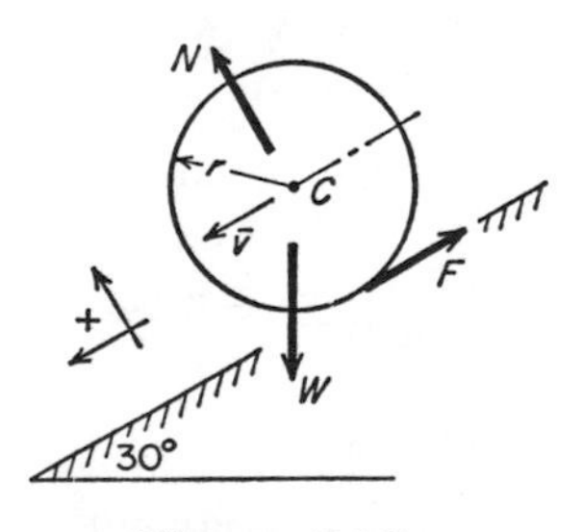

Figure 7.13

SOLUTION: Draw the free-body diagram, as in figure 7.13. Down the plane is positive.

$$\sum \mathbf{F}_x = W \sin 30^\circ - \mathbf{F} = \mathbf{R}_x$$

$$\mathbf{R}_x \Delta t = \frac{W}{g}(\overline{v}_2 - \overline{v}_1)$$

(a) $$W \sin 30^\circ - \mathbf{F} = \frac{W\overline{v}_2}{2g} \qquad \text{[FOR } \Delta t = 2\text{]}$$

Use $\overline{I} = 2mr^2/5 = \left(\frac{2}{5}\right)(W/g)r^2$ for a sphere and $\omega_2 = \overline{v}_2/r$ in the angular impulse and momentum equation ($\omega_1 = 0$):

$$\left(\sum \overline{M}\right) \Delta t = \mathbf{F} r \, \Delta t = \frac{2W}{5g} r^2 \frac{\overline{v}_2}{r}$$

$$\mathbf{F} = \frac{W\overline{v}_2}{5g} \qquad \text{[FOR } \Delta t = 2\text{]}$$

This value of $\mathbf{F}$ in equation (a) gives

$$W \sin 30 - \frac{W\overline{v}_2}{5g} = \frac{W\overline{v}_2}{2g}$$

$$\overline{v}_2 = 23 \text{ ft/sec} \quad (\textit{Ans.})$$

Law of Conservation of Momentum

If a body or a system of bodies is *not* acted upon by a net external force in a particular direction, the linear momentum of the body or system in that direction remains unchanged. It follows that if there is no resultant force ($\mathbf{R} = 0$), the linear momentum in any direction remains constant.

$$\mathbf{R}_x \Delta t = 0 = mv_2 - mv_1$$

$$mv_2 = mv_1 = mv_x$$

If two translating bodies of masses w_A lbm and w_B lbm have an impact, then during the time interval that no external forces act (the mass unit cancels),

$$w_A v_{A1} + w_B v_{B1} = w_A v_{A2} + w_B v_{B2}$$

[SUM TAKEN IN SOME x DIRECTION]

Angular momentum is conserved (remains constant) if the external moment M_o about a particular axis on a body or system of bodies is zero.

$$M_o \Delta t = 0$$

Hence,

$$\Delta I_o \omega = 0$$

$$I_o \omega = \text{a constant}$$

Impact

Direct impact occurs when the velocities of the two colliding bodies are each directed normal to the surfaces at the point of impact. *Central impact* occurs when the force of impact between two bodies is along the line joining the centers of gravity of the bodies. The velocities of bodies after impact depend on the law of the conservation of momentum and on the *coefficient of restitution*, which is defined by

$$e = \frac{\text{relative velocity of separation}}{\text{relative velocity of approach}} = \frac{v_{B2} - v_{A2}}{v_{A1} - v_{B1}}$$

$$e = -\frac{v_{B2} - v_{A2}}{v_{B1} - v_{A1}} = -\frac{v_{A2} - v_{B2}}{v_{A1} - v_{B1}}$$

The terms are algebraic, since the vectors are collinear. For actual bodies,

$$0 \leq e < 1$$

For any value of e less than unity, there will *always* be a loss of kinetic energy as a result of the impact, even though momentum *is* conserved. If there is any force acting after the impact, such as friction, momentum is being dissipated. For the *instant of impact*, it might be assumed that the momentum is conserved.

EXAMPLE: Mass A, 1 lbm, moving at a speed of 5 ft/sec, collides with mass B, 2 lbm, moving in the same direction at a speed of 2 ft/sec. Find the velocities of A and B after impact if the collision is perfectly elastic.

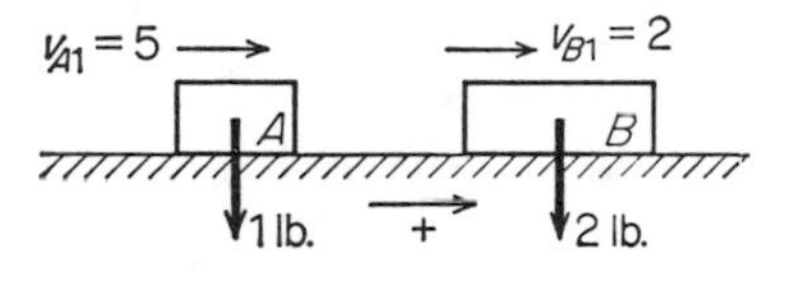

Figure 7.14

SOLUTION: Assume a direct central impact under frictionless conditions. For $g = g_0$, weight in lbf = mass in lbm. In figure 7.14, the bodies are shown moving in the positive x direction. (The figure is not a free-body, but would be if the normal forces were added.) Perfectly elastic bodies have a coefficient of restitution of $e = 1$.

$$e = 1 = -\frac{v_{B2} - v_{A2}}{v_{B1} - v_{A1}} = \frac{v_{A2} - v_{B2}}{2 - 5}$$

$$v_{A2} = v_{B2} - 3$$

Use this in the conservation of momentum (because no forces are acting in the x direction):

$$w_A v_{A1} + w_B v_{B1} = w_A v_{A2} + w_B v_{B2}$$
$$(1)(5) + (2)(2) = (1)(v_{B2} - 3) + 2v_{B2}$$
$$v_{B2} = +4 \text{ ft/sec, to right}$$
$$v_{A2} = 4 - 3 = +1 \text{ ft/sec, to right}$$

PRACTICE PROBLEMS

7.1. A mine hoist that, empty, weighs 5 tons is descending at a speed of 20 mph when the hoist mechanism jams and stops suddenly. If the cable is steel with a cross section of 1 in^2 and the hoist is 5000 feet below the hoisting drum, what is the stress produced in the cable?

Ans. approximately 54 ksi

7.2. A 2200–lbf automobile travels up a 3% incline at 45 mph. What horsepower is required?

Ans. 7.92 hp

7.3. A freight elevator weighs 1200 lbf. It hangs by a steel cable from a drum 5 feet in diameter. The drum, with shaft, weighs 2000 lbf. Its radius of gyration is $k = 1.8$ feet. In the usual method of operation, the elevator is allowed to drop freely until it reaches a speed of 20 ft/sec. There is 20 ft-lbf of torque on the drum shaft due to friction. How long does it take for the elevator to go from rest to 20 ft/sec? What is the tension in the cable?

Ans. 1.165 sec, 560 lbf

7.4. A cylinder weighing 200 lbm and having a 2 foot radius rolls, without slipping, on a horizontal surface. It is actuated by a cord that is wrapped around a concentric groove (radius = 1 foot) in the central plane of the cylinder. The cord comes out of the lower side of the groove, parallel to the horizontal plane to the right and 1 foot above the plane. A horizontal tension of 100 lbf is applied to the cord toward the right. Find the linear acceleration of the central axis of the cylinder. In which direction does the cylinder turn? (Neglect the groove in inertia calculations.)

Ans. 5.37 ft/sec^2, CW

7.5. A crate weighing 300 lbf slides from rest 15 feet down a chute inclined 30° with the horizontal floor. At this latter point (15 feet from starting), the chute turns to the horizontal plane along the floor. The crate slides along this horizontal section of the chute some distance before it comes to rest. The coefficient of friction is $f = 0.25$ between the crate and the chute. How far out along the horizontal section of the chute will the crate slide?

Ans. 17 feet

7.6. A nozzle discharges 40 gpm with a velocity of 60 ft/sec. Find the total force necessary to hold a flat plate in front of the nozzle perpendicular to the jet. Find the maximum pressure on the plate in psi.

Ans. 10.35 lbf, 24.2 psi

7.7. A particle of mass 2 slugs starts from rest at point A in figure 7.15 and slides down the side of a hemispherical surface of radius R under the action of gravity. If there is no friction, what force will be exerted on the surface by the particle at the instant it reaches B?

Ans. 193.2 lbf

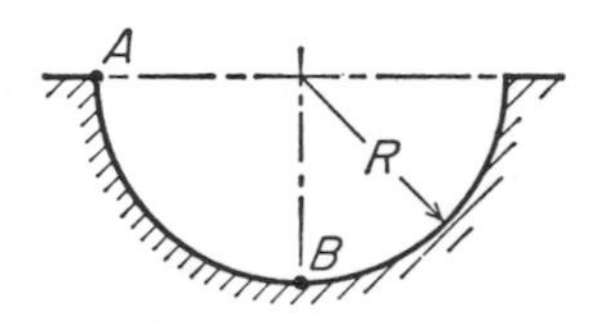

Figure 7.15

7.8. At the beginning of the drive, a golf ball has a velocity of 170 mph. If the club stays in contact with the ball for 1/25 second, what is the average force on the ball? The weight of the ball is 1.62 ounces.

Ans. 19.6 lbf

7.9. A container partially filled with water and having a total weight of 10 lbf is sliding down a 30° inclined plane where the coefficient of friction is 0.3. (a) What is the acceleration of the container? (b) What angle does the water surface in the container make with the horizontal.

Ans. (a) 7.72 ft/sec^2, (b) 13.5°

7.10. (a) Three springs, having spring constants of 5 lbf/in., 10 lbf/in., and 15 lbf/in., respectively, are connected in series as shown in figure 7.16. What is the magnitude of the combined spring constant? (b) If the springs were arranged in parallel in such a way that a load deflected them equally, what would be the magnitude of the deflection for a 100–lbf load?

Ans. (a) 2.72 lbf/in., (b) 3.33 inches

7.11. A car weighing 3100 lbf is slowed from 55 mph to 45 mph in 1.35 seconds, with maximum braking force applied. (a) If the same car were traveling at 65 mph, what length of time would be required to bring the car to a stop with maximum braking force applied? (b) What distance would be required to bring the car to a stop from 65 mph? (c) Compare the energy absorbed by the brakes in reducing the speed from 65 mph to

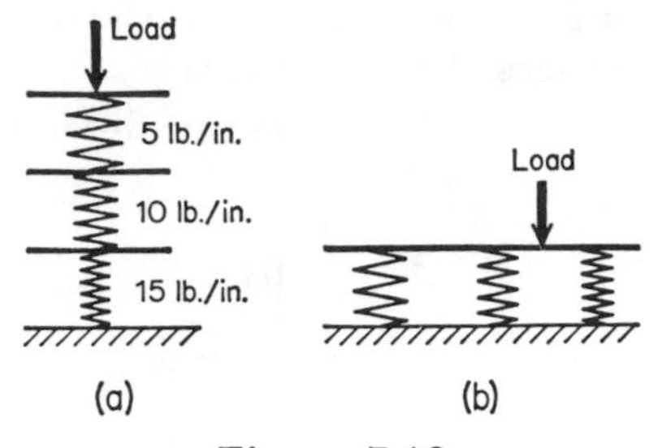

Figure 7.16

55 mph to that absorbed by the brakes in reducing the speed from 55 mph to 45 mph.
Ans. (a) 8.78 seconds, (b) 416 feet, (c) 123,600 ft-lbf versus 103,000 ft-lbf

7.12. Which of the following is a unit of energy? (a) hp, (b) watt, (c) ft-lbf, (d) ft/sec, (e) BTU/hr.

Ans. (c)

7.13. What average horsepower is required to lift an elevator weighing 3500 lbf, carrying a live load of 2000 lbf, from the first to the 20th floor (225 feet) of a building if the trip is made in 1 minute, 20 seconds? Show computations, neglecting friction and power required for acceleration and deceleration.

Ans. 28.1 hp

7.14. A gravel hoist for a concrete batching plant lifts gravel a total height of 35 feet at the rate of 25 buckets per minute. Each bucket contains 2.5 cu ft of gravel, with a specific weight of 105 lbf/ft^3. The hoist requires 1.5 hp to overcome friction. (a) What is the magnitude of the total horsepower required? (b) What is the magnitude of the torque required for a 1150-rpm motor?

Ans. (a) 8.47 hp, (b) 38.7 ft-lbf

7.15. A 3-ounce machine-gun bullet's exit velocity at the muzzle is 1500 ft/sec. Compute its kinetic energy at exit and its average acceleration while it is in the 30-inch long barrel.

Ans. 6550 ft-lbf, 450,000 ft/sec^2

7.16. In figure 7.17, a force of 2.5 lbf moves the block A horizontally with a constant velocity of 4 ft/sec. The coefficient of friction between blocks A and B is 0.10 and the friction of the rollers is considered negligible. Find

(a) the velocity of block B upwards, and (b) the force F_B exerted on block B to maintain the conditions of the problem.

Ans. (a) 3.0 ft/sec, (b) 2.72 lbf

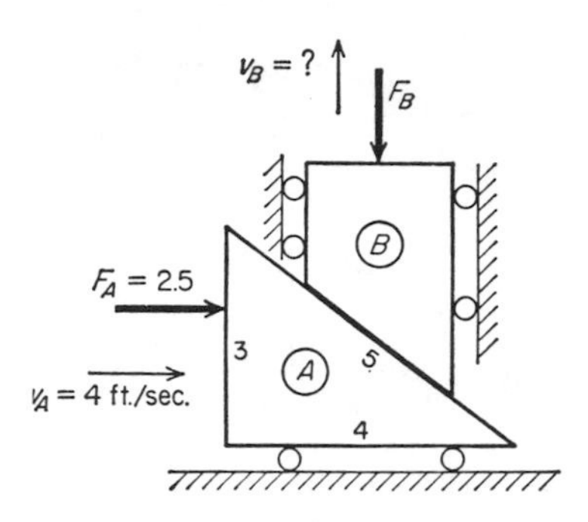

Figure 7.17

7.17. A packing box that weighs 64.4 lbf is delivered by a chute onto a horizontal table, as shown. It arrives on the table at point A with a horizontal velocity v. The coefficient of kinetic friction between the box and the table is constant. Assume that the box is of sufficiently large horizontal dimensions that it will not tip over. (a) If $v = 15$ ft/sec and the box comes to rest in 5 seconds after landing at A, find the coefficient of kinetic friction. (b) If $v = 10$ ft/sec and the box travels a distance of 14 feet after landing at A, find the coefficient of kinetic friction. (c) If the box comes to rest in 4 seconds after sliding a distance of 12 feet, find the coefficient of kinetic friction.

Ans. (a) 0.0931, (b) 0.111, (c) 0.0466

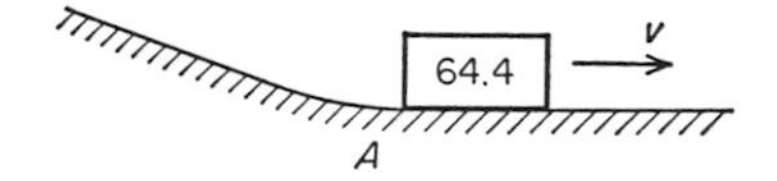

Figure 7.18

7.18. (a) State the relation between the work done on a body and the changes in its energy. (b) State Newton's three laws of motion relating to inertia, acceleration, and equilibrium.

7.19. The first derivative with respect to velocity of kinetic energy is (a) force, (b) power, (c) kinetic energy squared, (d) momentum, (e) acceleration, (f) none of these.

Ans. (d)

7.20. A 3-lbf force is applied to the rope of a pulley system attached to a 16-lbf weight. The 16-lbf weight is raised 1 foot while the 3-lbf force moves 8 feet. (a) What is the theoretical mechanical advantage? (b) What is the actual mechanical advantage? (c) What is the efficiency of the system?

Ans. (a) 8, (b) 5.33, (c) 66.7%

7.21. An 80-lbf wheel rotates at 90 rpm. Diameter is 18 inches. A brake shoe is applied at the rim with a force of 20 lbf. Coefficient of friction between wheel and shoe is 0.2. The radius of gyration is 6 inches. How many revolutions will the wheel make before it stops?

Ans. 1.465

7.22. A 10-ton electric truck is to accelerate at the rate of 3 mph/sec. Its rolling resistance is 25 lbf/ton. If the truck employs a gear reduction between the motor and the wheels of 9:1 and the wheels are 36 inches in diameter, what are the power and the torque required of the motor to accelerate the truck at 25 mph? Gears have 90% efficiency.

Ans. 221 hp, 552 ft-lbf

7.23. A flywheel has a weight of 2 tons, nearly all of which is in the rim at a distance of 3 feet from the axis of the wheel. If a motor starts the fly-wheel from rest and, in 10 seconds, has it revolving at the rate of one revolution in 2 seconds, at what average horsepower has the motor been working in this interval? Neglect friction.

Ans. 1.0 hp

7.24. When two elastic bodies (each moving with a high velocity) collide, which of the following laws or combination of laws can be used to solve for the resulting velocity (or velocities)? (a) conservation of momentum, (b) conservation of energy, (c) Dalton's law, (d) conservation of momentum and conservation of energy, (e) Avogadro's law.

Ans. (d)

7.25. An object of mass m is traveling with constant velocity when it makes an inelastic impact with a stationary mass, $m/2$. After the impact, the resultant velocity of the two masses is in the same direction as the velocity of the mass m before impact. (a) Does the momentum before impact equal the momentum after impact? (b) Is the kinetic energy of the masses before impact equal to the kinetic energy of the masses after impact? If not, what is the ratio of the kinetic energy of the masses before impact to that after impact? (c) Is the total energy of the system conserved?

Ans. (a) Yes, (b) No, 3:2, (c) Yes

7.26. A concrete batch plant has a conveyor that lifts sand a total distance of 20 feet at the rate of 36 buckets per minute. Each bucket holds 2.50 cu ft of sand. When there is no sand being conveyed, the conveyor motor runs at 1760 rpm and operates against a torque of 3 ft-lbf caused by the friction of the system. What is the total power required (in horsepower) to run the conveyor at the rate above if the sand being conveyed weighs 120 lbf/ft^3? Assume the friction is the same whether the system is loaded or empty.

Ans. 7.56 hp

7.27. A train is hauled up a grade of 30 feet per mile along a straight track by an engine that is assumed to exert a constant tractive effort of 16,000 lbf. If the weight of the engine and train is 600 tons, the grade is 1 mile long, the velocity of the train at the bottom of the grade is 10 mph, and the frictional resistance is constant at 10 lbf/ton, determine: (a) the velocity of the train at the top of the grade, (b) the time required to go up the grade, (c) the total work done by the engine on the grade, (d) the horsepower developed by the engine at the bottom and at the top, and (e) the horsepower necessary to keep the train moving up at a uniform 20 mph.

Ans. (a) 22.8 mph, (b) 220 sec, (c) 8.448 EE7 ft-lbf, (d) 426 hp; 972 hp, (e) 852 hp

7.28. Calculate the superelevation in feet/feet necessary on a curve of 6000 foot radius so that a frictional force in the radial direction will not be necessary to ensure equilibrium. The curve is to be designed for traffic moving at 60 mph.

Ans. 0.04 ft/ft

7.29. A 4800-lbm elevator starts from rest and is pulled upward with a constant $a = 10$ ft/sec^2. (a) Find the tension in the supporting cable. (b) What is the speed of the elevator after it has risen 45 feet? (c) Find the kinetic energy of the elevator 3 seconds after it starts. (d) How much is the potential energy of the elevator increased in the first 3 seconds? (e) What horsepower is required when the elevator is moving 22 ft/sec?

Ans. (a) 6290 lbf, (b) 30 ft/sec, (c) 67,050 ft-lbf, (d) 216,000 ft-lbf, (e) 252 hp

7.30. The flywheel of a punch press has a moment of inertia of 15 slug-ft^2 and operates at 300 rpm. It supplies the energy needed in a punch operation. Find the speed in rpm to which the wheel will be reduced in a sudden punching requiring 4500 ft-lbf of work.

Ans. 188 rpm

7.31. A 10-hp, 1750-rpm motor is connected directly to a 30 ft-lbf torque brake. The rotating system has a total inertia (Wr^2) of 10 lbf-ft^2. How long will it take the system to stop if the brake is set at the instant the motor is shut off? How much energy must be dissipated?

Ans. 1.9 sec, 5220 ft-lbf

7.32. A cable 1000 feet long and weighing 2 lbf/ft is suspended from a winding drum. It hangs vertically down a shaft. What work must be done against gravity to wind up the 1000 feet of cable on the drum?

Ans. 1 EE6 ft-lbf

7.33. An elevator weighing 2000 lbf is moving vertically upward with an acceleration of 3 ft/sec^2. A man standing in this elevator has an apparent weight of 180 lbf. The elevator is reversed and moves down with $a = 4$ ft/sec^2. Find the tension in the supporting cable under these conditions.

Ans. 2366 lbf, 1896 lbf

7.34. Compute the kinetic energy of a solid cast iron sphere 10 inches in diameter that is rolling freely along a horizontal plane at a velocity of 30 ft/sec. Cast iron weighs 450 lbf/ft^3.

Ans. 2673 ft-lbf

7.35. A motorcycle and rider weigh 400 lbf. They travel horizontally around the inside of a hollow vertical cylinder 100 feet in diameter. Find the minimum coefficient of friction that will allow this to be done at a speed of 40 mph.

Ans. 0.467

7.36. A weight of 50 lbf rests on a vertical coil spring whose deflection rate is 500 lbf/in. The spring and weight are then pressed down farther by an additional force of 3000 lbf. If this added force of 3000 lbf is suddenly released, what will be the speed of the 50-lbm weight at a point 4 feet above the depressed condition?

Ans. 27.2 ft/sec

7.37. A body weighing 50 lbf is placed on an inclined plane that makes an angle of 30° with the horizontal. The coefficient of friction is 0.12. Find the effective force down the plane and the acceleration. What would be the velocity at the end of 3 seconds?

Ans. 19.8 lbf, 12.75 ft/sec^2, 38.25 ft/sec

7.38. A train whose mass is 200 tons is moving on level track at a speed of 30 mph. When the power is shut off and brakes applied, there is a frictional resistance of 20 lbf/ton. Find how far from the station the power must be shut off and the brakes applied so that the train will stop at the proper place. How much work is done in bringing the train to a stop?

Ans. 3000 ft, 1.2 EE7 ft-lbf

7.39. A body that weighs 100 lbf starts from rest and slides down an inclined plane that is 40 feet long and inclined 22° with the horizontal. After leaving the inclined plane, the body continues to slide on a horizontal plane until it comes to rest. The coefficient of friction on the inclined plane is 0.2 and, on the horizontal, 0.3. Determine the acceleration of the body on each plane and the amount of work done on the body from the time it starts until it stops.

Ans. 6.1 ft/sec^2, 9.66 ft/sec^2, 1500 ft-lbf

7.40. A metal disk of uniform thickness is 6 feet in diameter and weighs 600 lbm. If the disk is rotating on its central axis at 240 rpm, what is its kinetic energy?

Ans. 26,500 ft-lbf

7.41. A 50-ton car moving at 3 mph strikes a bumping post equipped with a 40,000-lbf spring. If all the shock is absorbed by the spring, how far will the spring be compressed?

Ans. 4.25 inches

7.42. An elastic body weighing 100 lbm is moving with a velocity of 20 ft/sec and overtakes another elastic body weighing 50 lbm moving in the same direction at 15 ft/sec. What is the final velocity of each body, and what is the loss of kinetic energy?

Ans. 16.67, 21.67 ft/sec, $\Delta E_k = 0$

7.43. A standpipe is 50 feet high and 10 feet in diameter. It is filled by forcing water through the bottom from a pond whose level is maintained 40 feet below the bottom of the standpipe. Frictional resistance of the water in passing through the piping should be considered as equivalent to an additional lift of 10 feet. Assume an efficiency of 60% for the pumping unit and a time of 2 hours. What is the size of the motor and the capacity (quantity and head) of the pump?

Ans. 10.3 hp, 245 gpm, 100 ft

7.44. A water wheel makes 8 rpm. The tangential force of the water that operates the wheel is equal to $\frac{3}{4}$ ton at a radius of 12 inches. With an overall efficiency of 25.7%, what horsepower is delivered?

Ans. 8.87 hp

7.45. A 210-lbf man slides down a rope that has a breaking strength of 160 lbf. What is the man's least acceleration if the rope is not to break?

Ans. 7.67 ft/sec^2

7.46. A 2-ounce bullet with an initial velocity of 500 ft/sec is fired horizontally into a 50-lbm block of lead that rests on a frictionless horizontal surface. What is the speed of the block and bullet?

Ans. 1.246 ft/sec

8 *STRENGTH OF MATERIALS*

SYMBOLS

A = area
c = distance from neutral axis to extreme fiber
D = diameter
E = modulus of elasticity in tension or compression (σ/ϵ)
F = force; total load
f = stress symbol used in reinforced concrete
G = modulus of elasticity in shear or torsion
I = rectangular moment of inertia of area about centroid
J = polar moment of inertia of area about centroid
k = radius of gyration
L = length
M = moment of force; bending moment at a section
m = mass in slugs
N = factor of safety or design factor; quantity
n = rpm; a ratio of moduli of elasticity
p = pressure, psi
R = resultant force
r = radius
S = strength
T = torque
V = vertical shear, beam
W = total load or force of gravity
w = load per unit length (lbf/ft or lbf/in)
y = deflection of beam

Z = section modulus (I/c)
α = coefficient of thermal expansion (linear)
δ = total strain ($L\epsilon$)
ϵ = unit strain (σ/E)
θ = angular deflection in torsion, radians
σ = resultant normal stress in combined stresses
σ_c = compressive stress
σ_f = flexural stress
σ_t = tensile stress
τ = resultant shear stree in combined stresses

This phase of applied physics is also called *mechanics of materials*. You might be expected to know:

for all steels,

$$E \approx 3 \text{ EE7 psi} = 3 \text{ EE4 ksi}$$
$$G \approx 1.2 \text{ EE7 psi} = 1.2 \text{ EE4 ksi}$$
$$\text{endurance limit } S_e \approx \frac{S_u}{2}$$
$$\text{ksi} = \text{kips per in}^2$$

Definitions

In this area, the basic units generally are length in inches and force in pounds (or kips to reduce the size of the number). The *ultimate strength*, S_u, is the maximum tensile load on a specimen tested to rupture divided by the original area. The *yield strength*, S_y, is the stress in the tensile specimen under test when there is a marked deformation (yielding) without much change in load, or it is arbitrarily where a 0.2 percent (usually) offset line intersects the σ-ϵ curve. The *endurance limit*, S_e is the maximum reversed stress that can be repeated an indefinite number of times on a standard specimen in bending without causing failure. These strengths, all of which are test values available in handbooks and texts, are criteria for design.

In order for a design to be safe, these stresses must be modified by a so-called *factor of safety* (preferably called a *design factor*). One of these stresses divided by a factor of safety gives a number variously called *design stress* (best), *working stress, safe stress, allowable stress*, and even *induced stress* (not good in this context). Factors of safety (and design stresses) used in particular designs depend variously upon experience (a broad term), the strength criterion to be used, the theory and method of design, codes that specify design stresses, and the mood and judgment of the engineer who makes the decision. In some situations, the stress involved is not a significant parameter. Rather, the design must be made on the amount of permissible deformation.

SIMPLE STRESSES

For tension (line of action of F through centroid),

$$\sigma_t = F/A \text{ psi}$$

$$\text{deformation } \delta = \epsilon L = \frac{\sigma_t L}{E} = \frac{FL}{AE} \text{ inches}$$

For compression (line of action through centroid),

$$\sigma_c = F/A \text{ psi}$$

$$\text{deformation } \delta = \epsilon L = \frac{\sigma_c L}{E} = \frac{FL}{AE} \text{ inches}$$

For shear (load parallel to area),

$$\tau_s = F/A \text{ psi}$$

If a body is prevented from expanding or contracting when its temperature changes, a stress is introduced. Some coefficients of linear expansion at normal temperatures are:

steel,	$\alpha = 6$ EE–6 in/in-°F
concrete,	$\alpha = 6.2$ EE–6 in/in-°F
yellow brass,	$\alpha = 1.13$ EE–5 in/in-°F
aluminum,	$\alpha = 1.29$ EE–5 in/in-°F

Free expansion with temperature change $= \Delta T$ is

$$\Delta L = \alpha L \Delta T \text{ inches}$$

L inches equals length and ΔT°F equals change in temperature.

Example: A slab of concrete is 25 feet long at 30°F. (a) How much longer will it be at 120°F if it is free to expand? (b) If the slab is constrained to allow only one-third of the free expansion, what compressive stress is developed? The modulus of elasticity for this concrete is 2.5 EE6 psi.

SOLUTION:

(a) $\Delta L = (6.2\ \text{EE–6})(25 \times 12)(120 - 30) = 0.1673\ \text{inches} \quad (\textit{Ans.})$

(b)
$$\text{effective deformation } \delta = \frac{2}{3}(0.1673) = 0.11153 \text{ inches}$$
$$\text{unit deformation} = \epsilon = 0.11153/300 \text{ in/in}$$
$$\text{stress} = \sigma_c = \epsilon E = \frac{0.11153}{300} \times 2.5 \text{ EE6}$$
$$= 930 \text{ psi} \quad (\textit{Ans.})$$

BEAMS

The bending stress in a straight beam at a point where the bending moment is M is

$$\sigma_f = \frac{Mc}{I}$$

In the above equation, I is the centroidal moment of inertia and c is the distance from the neutral plane axis to the most extreme fibers in bending.

If the section is symmetric, the *section modulus* $Z = I/c$ in^3 is convenient:

$$\sigma_f = M/Z$$

The *moment at a section* is the sum of the moments of *all* the external forces to the right (or left) of the section about an axis through the centroid and perpendicular to the beam's length. (Make a free-body diagram.) The *maximum moment* on a beam occurs where the shear diagram crosses zero. If it crosses more than once, compute M at each such point to determine the maximum. Handbooks give the maximum moments for some simple beam arrangements. Bending stress σ_f is either tension (on one side of neutral plane) or compression (on the other side). For bending on areas nonsymmetric about the neutral axis (but symmetric about the perpendicular centroidal axis),

$$\sigma_t = \frac{Mc_t}{I}$$
$$\sigma_c = \frac{Mc_c}{I}$$

c_t is the distance from the neutral plane to the extreme fiber on the tensile side. c_c is the analogous distance on the compressive side.

Vertical Shear

Vertical shear, V lbf, at a section in a beam is the sum of the vertical forces (i.e., perpendicular to the length) either to the right or to the left of the section. If a beam with vertical forces is cut into two parts, each part can be placed in equilibrium by inserting on the section a force V and a moment M in the proper senses. The magnitude of the vertical shear stress (equal to the horizontal shear stress) at a distance c from the neutral axis (rectangular shapes) is given by

$$\tau_s = \frac{V}{Ib'} A'\overline{y}$$

A' is the shaded area in figure 8.1, and $A'\overline{y}$ is the moment of this area about the neutral axis. I is the centroidal moment of inertia of the whole section. b' is the width of section at the point where the stress is desired. In finding the moment of A', take it as a composite area composed of two rectangles *efgh* and *kmxx* and use the principle

$$A'\overline{y} = A'_1 y'_1 + A'_2 y'_2$$

Handbooks give maximum values (at the neutral axis) of the vertical shear stress for some simple cases.

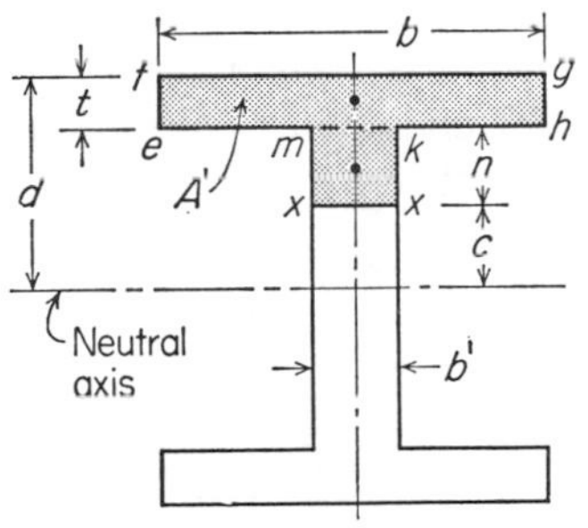

Figure 8.1

$$V = \frac{dM}{dx}$$

$$\Delta M = M_A - M_B = \int V dx$$

ΔM is the change of moment between sections (say, A and B) and $\int V dx$ is recognized as the area of the shear diagram between these sections. Deformations of beams require extended computations, except in the simple cases that can be found in tables. Shear and moment diagrams for some simple situations are shown in figure 8.2: (a) is a cantilever with concentrated load; (b) is a cantilever with uniformly distributed load; (c) is a simple beam with load at center; and (d) is a simple beam with uniformly distributed load.

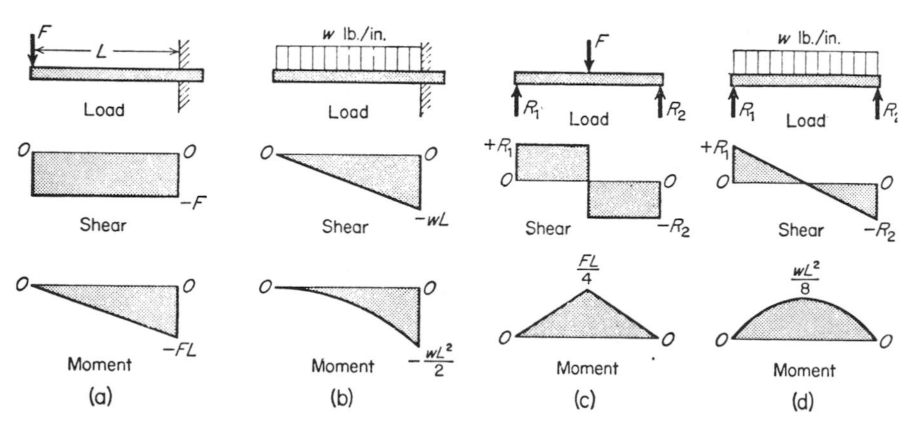

Figure 8.2

EXAMPLE 1: A simple beam 16 feet long has two supports 12 feet on centers with one at the left end. The beam is uniformly loaded with 100 lbf/ft. Find the reactions at the supports, the maximum moment, and the maximum value of the shear.

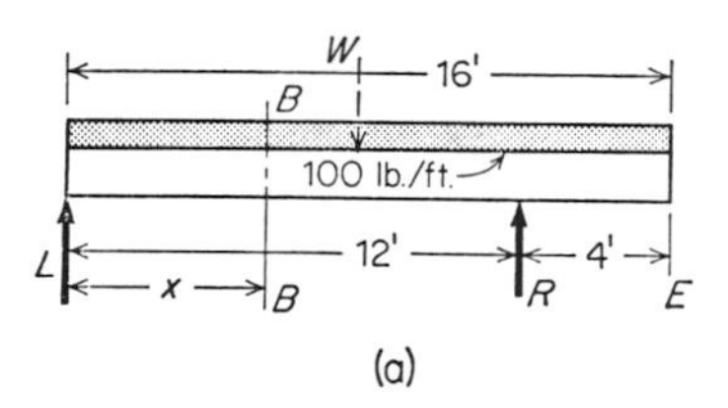

Figure 8.3 (a)

SOLUTION: Sketch the beam as in figure 8.3(a), and then any free-bodies as needed. For finding the reactions, note the symmetry of loading. The equivalent $W = (16)(100) = 1600$ lbf is at the midpoint of beam. This force can be used only for finding the external reactions; the internal forces (stresses) are affected by the manner in which the external loads are applied. Then, referring to figure 8.3(a),

$$\sum M_L = 12R - (1600)(8) = 0$$
$$R = 1066.7 \text{ lbf} \quad (\textit{Ans.})$$

$$\sum M_R = 12L - (1600)(4) = 0$$
$$L = 533.3 \text{ lbf} \quad (\textit{Ans.})$$

$$\sum F_y = 533.3 - 1600 + 1066.7 = 0 \quad \text{(Check)}$$

The moment at any section between L and R at x feet from left end is obtained from the free-body diagram of a section, figure 8.3(b).

$$\sum M_B = M_B = (533.3)(x) - (100x)\frac{x}{2} = 533.3x - 50x^2$$

The shear at any section is obtained from $\sum Fy$ or from $V = dM/dx$.

$$\sum F_y = 533.3 - 100x - V_B = 0$$
$$V_B = 533.3 - 100x = \frac{dM}{dx}$$

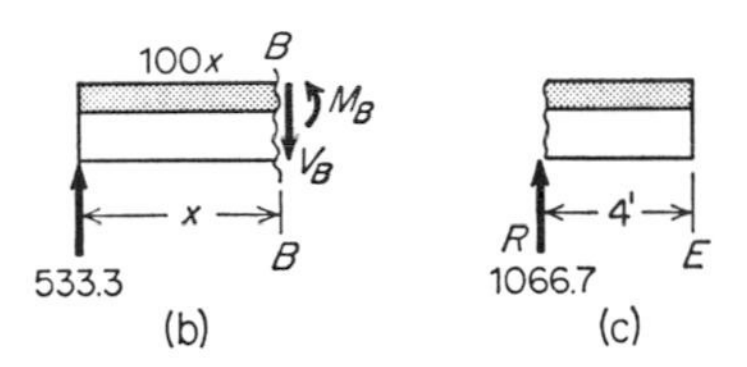

Figure 8.3 (b) and (c)

The senses of the resisting shear and moment at any section B are opposite to the resultant external shear and moment on the section. The moment is a maximum when $dM/dx = V = 0$, or at $x = 5.333$ feet. From the moment equation,

$$\begin{aligned} \text{maximum } M_B &= (533.3)(5.333) - (5)(5.333)^2 \\ &= 1420 \text{ ft-lbf} = 17{,}400 \text{ in-lbf} \quad (\textit{Ans.}) \end{aligned}$$

The moment is discontinuous at R, but a quick check of the right-hand end as a free-body, as in figure 8.3(c), shows that M_{max} does not occur there. Maximum shear nearly always occurs at a point of application of a load, L or R in this beam. The safe way of being sure of maximums is to sketch shear and moment diagrams.

$$\text{At } L\ (x=0),\ V = 533.3 \text{ lbf}$$

between L and R

$$\text{At } R\ (x = 12 - dx),\ V = -666.7 \text{ lbf}$$
$$\text{At } R\ (x = 12 + dx),\ V = -666.7 + R = -666.7 + 1066.7 = 400 \text{ lbf}$$

$$\text{maximum shear } V = 666.7 \text{ lbf at } R \quad (Ans.)$$

The question might require you to sketch moment and shear diagrams, which is the safe thing to do anyway. From the moment equation, M_B is zero at $x = 0$ and $x = 10.67$ feet from the left. At R, $M_R = (100)(4)(2) =$ 800 ft-lbf. At E, $M = 0$. $M_{\text{max}} = 1420$ ft-lbf at $x = 5.333$ feet. The curve, as seen from its equation, is a parabola. Spot these points and sketch the curve, as in figure 8.4. The vertical shear curve is sketched from data previously calculated. Unless otherwise required, freehand sketches with regard for proportion usually serve the purpose of aiding and checking solutions. The part of the area of the shear diagram that lies above the axis $V = 0$ is equal to the part that lies below.

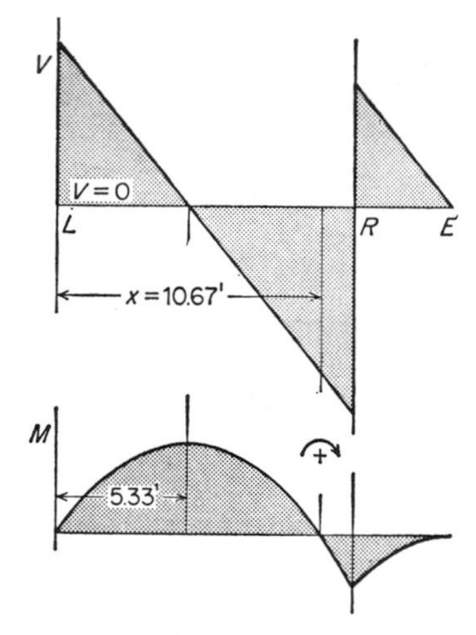

Figure 8.4

Deflection

Various methods are used. The basic equation is

$$EI\frac{d^2y}{dx^2} = M$$

EXAMPLE 2: Find the maximum deflection of a cantilever beam with a uniform load of w lbf/in.

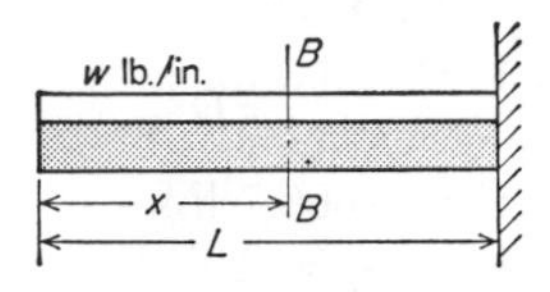

Figure 8.5

SOLUTION: First, set up an expression for M at any section as in figure 8.5. Then, integrate.

$$M = M_B = (wx)\frac{x}{2} = \frac{wx^2}{2} = EI\frac{d^2y}{dx^2}$$

$$EI\frac{dy}{dx} = \frac{wx^3}{6} + C$$

When $dy/dx = 0$, $x = L$, and $C = -wL^3/6$. Using this value of C and integrating again,

$$EIy = \frac{wx^4}{24} - \frac{wL^3}{6}x + C'$$

When $y = 0$, $x = L$, and $C' = wL^4/6 - wL^4/24 = wL^4/8$,

$$EIy = \frac{wx^4}{24} - \frac{wL^3}{6}x + \frac{wL^4}{8}$$

y is the deflection at any section x inches from the free end. y_{max} is at the end where $x = 0$, or

$$y_{\text{max}} = \frac{wL^4}{8EI} = \frac{WL^3}{8EI} \quad (\textit{Ans.})$$

E is in psi, I is in in^4, L is in inches, w is in lbf/in, and y is in inches. For all beams in bending under a single load F or a total distributed load W, the deflection is CFL^3/EI or CWL^3/EI, where C is a constant for a particular type of beam.

TORSION

Torsion, at a section where the applied torque or twisting moment is T in-lbf, is $\tau_s = Tc/J$ psi where J in^4 is the centroidal polar moment of inertia of the *section*, and c is in inches as for bending. Strictly, this is good only for round sections. Let $Z' = J/c$ in^3, then $\tau_s = T/Z'$. For a *solid* round section,

$$J = \frac{\pi D^4}{32}$$
$$Z' = \frac{\pi D^3}{16}$$
$$\tau_s = \frac{16T}{\pi D^3}$$

Angular deflection of a section under torque T in-lbf, *round* member, is

$$\theta = \frac{TL}{GJ} \text{ radians}$$

L inches is the length of the member *between* the section where T is applied and the "fixed" section.

EXAMPLE: A 2-inch solid shaft is driven by a 40-inch gear and transmits power at 100 rpm. If the allowable shearing stress is 12 ksi, what horsepower can be transmitted? See figure 8.6.

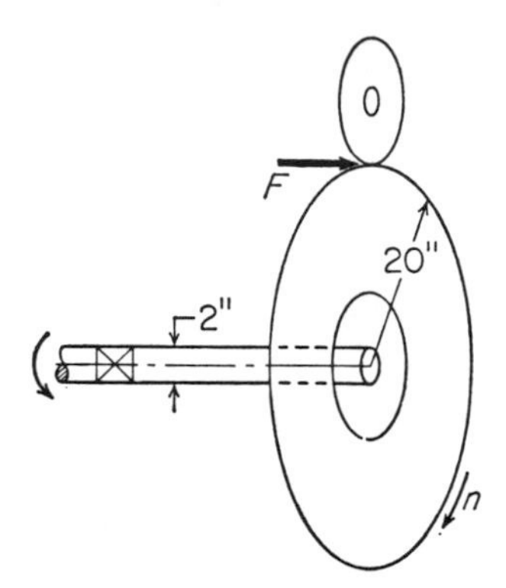

Figure 8.6

SOLUTION: The allowable torque is

$$T = \sigma_s \frac{\pi D^3}{16} = \frac{(12{,}000)(\pi 8)}{16} = 18{,}850 \text{ in-lbf}$$

The force at the pitch line is $F = T/r = 18{,}850/20 = 942.5$ lbf. Work of F is $Fv = F(2\pi rn)$ ft-lbf/minute and horsepower is

$$\text{hp} = \frac{(942.5)(2\pi 20/12)(100)}{33{,}000} = 29.9 \text{ hp} \quad (\textit{Ans.})$$

COLUMNS

For long columns, Euler's equation is theoretically valid:

$$F_c = \frac{C\pi^2 EA}{(L/k)^2} = NF$$

F_c is the critical or buckling load, N is a factor of safety for an actual load F, C is a factor that is a function of the end restraints (usually C is taken as 1 except for a free-end column whose value of $C = \frac{1}{4}$), E psi is the modulus of elasticity, A in^2 is the section area, L inches is the length of the column, and k inches is the *least* radius of gyration. L/k is called the *slenderness ratio.*

For very short compression members, there is no buckling and $F = \sigma A$. For intermediate lengths, various empirical formulas are generally used: straight line formulas, Rankine, or Johnson. Johnson's formula (also called *parabolic formula*) is

$$\frac{F}{A} = \sigma_e \left[1 - \frac{\sigma_y (L/k)^2}{4C\pi^2 E}\right] \qquad \left[40 < \frac{L}{k} < 120, \text{ MILD STEEL}\right]$$

The Rankine formula is

$$\frac{F}{A} = \frac{a}{1 + \frac{1}{b}\left(\frac{L}{k}\right)^2}$$

$\sigma_e = S_y/N$ is an *equivalent stress*, which is some safe stress in design. Other symbols have their usual meanings and units. Handbooks give values of a and b, which depend on the mechanical properties of the material and design factors, but, for structures, they are specified in codes in forms such as

$$\text{straight-line formula:} \quad \frac{F}{A} = a + b\frac{L}{k}$$

$$\text{parabolic formula:} \quad \frac{F}{A} = a + b\left(\frac{L}{k}\right)^2$$

ECCENTRIC LOADING

If a tensile member supports an eccentric load, as in figure 8.7(a), the resultant maximum normal stress is

$$\sigma = \sigma_1 \pm \sigma_2 = \frac{F}{A} \pm \frac{Fec}{I}$$

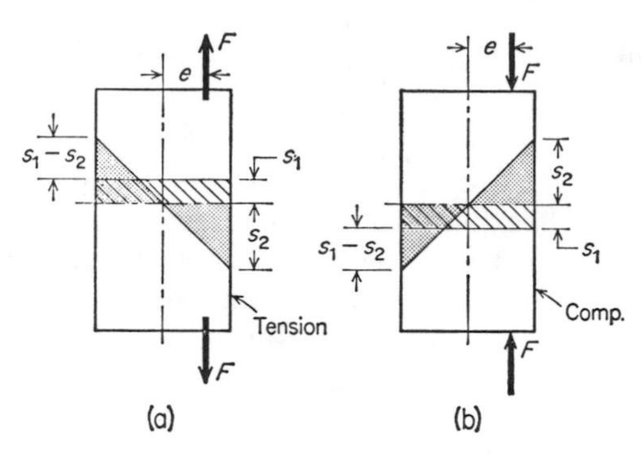

Figure 8.7

$Fe = M$ = the bending moment existing because of the load's eccentricity, the plus sign applies to the side toward which the load is located (from the center line), and I/c is as usual. If the eccentric load is compressive, as in figure 8.7(b),

$$\sigma = -\sigma_1 \pm \sigma_2 = -\alpha\frac{F}{A} \pm \frac{Fec}{I}$$

The negative signs indicate a compressive stress, and the maximum normal stress is $\sigma = -\sigma_1 - \sigma_2$ on the side of the eccentricity. α is introduced to care for column action and is taken as unity if the column is short, say $L/k < 40$. The foregoing equations are applied to unsymmetric sections (I always with respect to centroidal axis), but are inaccurate on the unsafe side unless the member is straight (in direction of the load).

Any area subjected to both a bending stress Mc/I from any cause and a normal F/A stress is treated as described above. At a point in the area,

$$\sigma = \frac{F}{A} \pm \frac{Mc}{I}$$

EXAMPLE: A 4-foot square concrete pier is to carry a load of 576 kips. Compare the greatest stress in the pier when the load is applied on one center line but 4 inches from the other center line with the stress that would exist if the load were centrally applied so that $\sigma = P/A$. Choose your answer from the following: (a) $1.5P/A$, (b) $3P/A$, (c) $2P/A$, (d) $2.5P/A$.

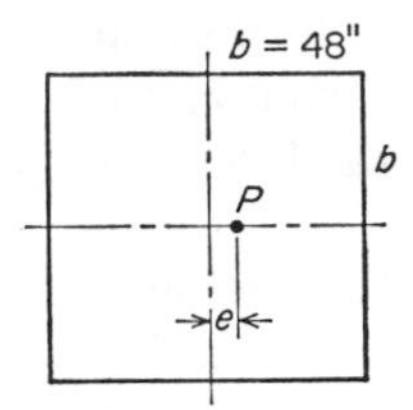

Figure 8.8

SOLUTION: Figure 8.8 is a view down onto the top of the pier.

$$e = 4'' = 4b/48$$
$$I = b^4/12$$
$$c = b/2$$
$$A = b^2$$
$$\sigma = \frac{P}{A} + \frac{Pec}{I} = \frac{P}{A} + P\left(\frac{4b}{48}\right)\left(\frac{b/2}{b^4/12}\right)$$
$$= \frac{P}{A} + \frac{P}{2b^2} = \frac{P}{A} + \frac{P}{2A} = 1.5\frac{P}{A} \quad \text{(a)} \quad (\textit{Ans.})$$

COMBINED NORMAL AND SHEARING STRESSES

If a point in a section has on it a single normal stress σ (either tension or compression) and a shear stress τ_s, the principal normal stresses in a plane are

$$\sigma_1, \sigma_2 = \frac{\sigma}{2} \pm \sqrt{\tau_s^2 + \left(\frac{\sigma}{2}\right)^2}$$

This equation is the mathematical expression of the *maximum principal stress theory* and is recommended for brittle materials. The positive sign gives the maximum stress. If the sign of the answer is positive, σ is the same kind of stress as σ (tension or compression). If the sign of the answer is negative, the stress σ is the minimum principal stress in this plane and is compression if σ is tension, tension if σ is compression.

The maximum shear stress for the *point* and section as described above is

$$\tau_{\text{max}} = \frac{1}{2}(\sigma_{\text{max}} - \sigma_{\text{min}}) = \sqrt{\tau_s^2 + \left(\frac{\sigma}{2}\right)^2}$$

This is the mathematical expression of the *maximum shear stress theory.* In each of the foregoing equations, the normal stress σ is computed from F/A or Mc/I, and the shear stress is computed from F/A or Tc/J. Experience is necessary for proper use of the foregoing equations.

EXAMPLE: A screw jack has an axial load $W = 8000$ lbf located slightly off center at an eccentricity of $e = 0.25$ inches. To turn the screw, a force of $F = 60$ lbf is applied at a radius of 20 inches from the axis of the screw and at the top when the screw is extended 12 inches. Assume there is no tendency to buckle. Other data are: $A = 2.4\ \text{in}^2$, $Z = 0.525\ \text{in}^3$, and $Z' = 2Z = 1.05\ \text{in}^3$. Compute the maximum shear stress if the loaded end moves only slightly as the screw bends.

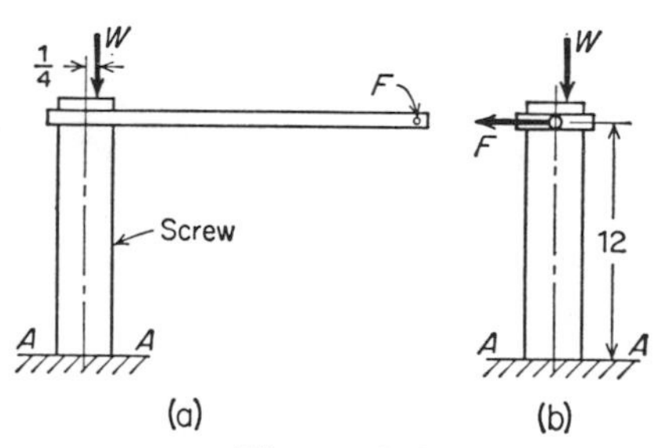

Figure 8.9

SOLUTION: Figure 8.9(a) is a diagrammatic representation that shows the point of application of F. After a 90° turn, the appearance might be somewhat as in figure 8.9(b). The maximum stress occurs in a section AA where the screw enters the base, as the following analysis shows. Uniform stress is

$$\sigma_1 = \frac{W}{A} = \frac{8000}{2.4} = 3330 \text{ psi}$$

This is the compressive stress on all sections from W to AA.

Bending by W is

$$\sigma_2 = \frac{We}{Z} = \frac{(8000)(0.25)}{0.525} = 3810 \text{ psi}$$

This is on all sections (extreme fibers) from W to AA.

Bending by F is

$$\sigma_3 = \frac{M}{Z} = \frac{(60)(12)}{0.525} = 1370 \text{ psi}$$

This is on section AA only (extreme fibers). Note in figure 8.9(b) that the screw is acting as a 12-inch cantilever.

At some stage of a rotation of the screw, all the normal compressive stresses act together at a particular point. Thus, the total normal stress is

$$\sigma_c = \sigma_1 + \sigma_2 + \sigma_3 = 3330 + 3810 + 1370 = 8510 \text{ psi}$$

The torsional stress all around the screw on outer fiber is

$$\tau_s = \frac{T}{Z'} = \frac{(60)(20)}{1.05} = 1140 \text{ psi}$$

This is on all sections from F to AA. The maximum shearing stress is

$$\tau_{\max} = \sqrt{\tau_s^2 + (\sigma/2)^2} = \sqrt{(1140)^2 + (8510/2)^2} = 4400 \text{ psi} \quad (Ans.)$$

Shear Stresses in Two or More Directions

There are two ways of handling coplanar shearing stresses: (a) Find shear stress from each force on the section and add vectorially, or (b) find the resultant *shearing* force on the section $R = \sum F$ (vector sum) and then the resultant stress $\tau_s = R/A$.

COMPOSITE BEAMS

Figure 8.10(a) shows a wood beam bonded to steel plates on the outside with neutral axis as shown. It is assumed that at any point P on the common surfaces, the unit deformations are the same in the steel and wood:

$$\epsilon_s = \epsilon_w = \frac{\sigma_s}{E_s} = \frac{\sigma_w}{E_w} = \frac{F}{A_s E_s} = \frac{F}{A_w E_w}$$

$$\frac{E_s}{E_\psi} = n = \frac{A_w}{A_s}$$

$$A_w = nA_s$$

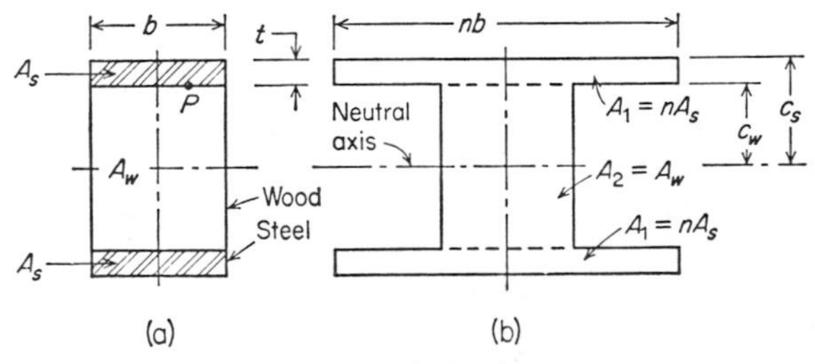

Figure 8.10

This means that the area of wood needed to be equivalent to the area of steel is $n = E_s/E_w$ times the area of steel. Make an equivalent wood beam, as in figure 8.10(b), in which $A_1 = nA_s$, keeping the same t. Then, treat the equivalent or transformed beam as homogeneous. Determine the centroidal I for the section in (b), treating as composite areas A_1 and A_2. Determine moment M at a section as usual. Compute the maximum stress in the wood from

$$\sigma_w = \frac{Mc_w}{I}$$

Compute the maximum stress in the steel from

$$\sigma_s = \frac{nMc_s}{I}$$

The modulus of elasticity in tension for wood might be about 1 EE6 psi to 1.5 EE6 psi.

PRACTICE PROBLEMS

8.1. A steel tie rod on a bridge must be made to withstand a pull of 5000 lbf. Find the diameter of the rod. Assume a factor of safety of 5 based on the ultimate stress of 64,000 psi.

Ans. 0.71 inches

8.2. A steel pipe having a wall thickness of $\frac{1}{4}$ inch and an inner diameter of 10 inches is subjected to steam pressure that causes a tensile unit stress of 20,000 psi on a longitudinal cross section of the pipe wall. Calculate the steam pressure.

Ans. 1000 psi

8.3. A hoisting cable $\frac{3}{8}$ inch in diameter supports a mine hopper that weighs 1000 lbf. The length of the cable when carrying the hopper is 1650 feet and the length of the cable when supporting the hopper and 1000 lbf of ore is 6.25 inches greater than when supporting the hopper alone. Find the modulus of elasticity of the cable.

Ans. 2.86 EE7 psi

8.4. If the ultimate shearing strength of a steel plate is 42,000 psi, what force is necessary to punch a $\frac{3}{4}$-inch diameter hole in a $\frac{5}{8}$-inch thick plate?

Ans. 61,800 lbf

8.5. The section modulus of a uniform beam means: (a) the density of the beam at the section under consideration, (b) the moment of inertia of the section about its neutral axis divided by the distance from the neutral axis to the outermost fiber, (c) the weight of the beam in pounds per foot of length, (d) the moment of inertia of the beam section about its neutral axis divided by the weight per foot, (e) none of these.

Ans. (b)

8.6. The yield strength of common yellow brass (70% Cu, 30% Zn) can be increased by: (a) heat treatment, (b) annealing, (c) chill casting, (d) cold working, (e) none of these.

Ans. (d)

8.7. A steel axle under repeated stress will eventually fail if that stress is above the endurance limit for the steel under consideration. The endurance limit of the steel is: (a) equal to the stress at the yield point, (b) equal to the ultimate tensile strength, (c) roughly equal to half the ultimate tensile strength, (d) roughly equal to half the elastic limit, (e) none of these.

Ans. (c)

8.8. From compressive tests of concrete cylinders, it has been found that the concrete fails when the unit deformation is about 0.0012. (a) How much does a specimen 8 inches in diameter by 16 inches high shorten before failure occurs? (b) For E_c = 2.5 EE6 psi, what is the stress at failure?

Ans. (a) 0.0192 inches, (b) 3000 psi

8.9. A steel shaft 4 inches in diameter is used as a cantilever beam and loaded as shown in figure 8.11. Find the maximum stress in the section A at the wall.

Ans. 6880 psi

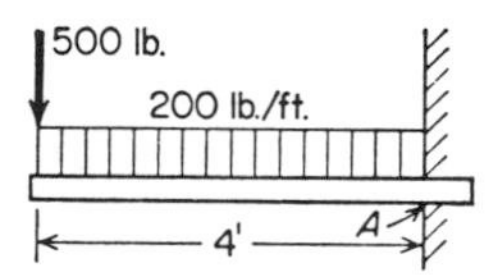

Figure 8.11

8.10. A straight bar 6 feet long and $\frac{7}{8}$ inch in diameter is turned down to a diameter of $\frac{1}{2}$ inch for a distance of 2 feet in its central portion. An axial load P causes a unit deformation of 0.001 in the central 2 feet, and a total stretch of 0.04 inches in the whole bar. What is the unit deformation of each of the end portions?

Ans. 3.33 EE–4

8.11. A bar of steel that has a diameter of 2 inches supports an axial tensile load P. It is observed that a 10-inch gauge length of the bar stretches 0.005 inch when the tensile unit stress increases from 15,000 to 30,000 psi. Calculate the modulus of elasticity of the material.

Ans. 3.0 EE7 psi

8.12. It is specified that a steel rod 100 inches long is to be subjected to a unit stress not greater than 10,000 psi and to be elongated not more than 0.01 inch when resisting a tensile axial load of 30,000 lbf. Determine the cross-sectional area required to satisfy each of the specifications, and state which requirement governs the design.

Ans. 3 in^2, 10 in^2; the latter

8.13. A simple beam (see figure 8.12) having a rectangular cross section 6 inches wide and 12 inches deep is subjected to a uniformly distributed load of 400 lbf/ft (including the weight of the beam) over the whole span, and a concentrated load of 2000 lbf at a distance of 4 feet from the left support. Find the tensile unit stress on the outer fiber of the beam at 5 feet from the left end.

Ans. 1370 psi

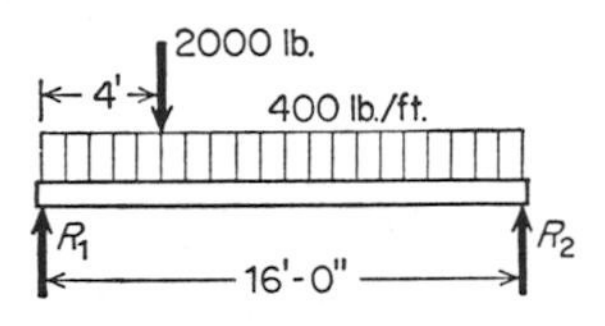

Figure 8.12

8.14. A rectangular wooden beam is loaded as shown in figure 8.13. (a) If the allowable shearing stress parallel to the grain is 68 psi, is the maximum shearing stress in the beam within the allowable limit? (b) If the allowable bending stress is 720 psi, is the bending stress in the beam within the allowable limit?

Ans. (a) No, 93.8 psi, (b) No, 1685 psi

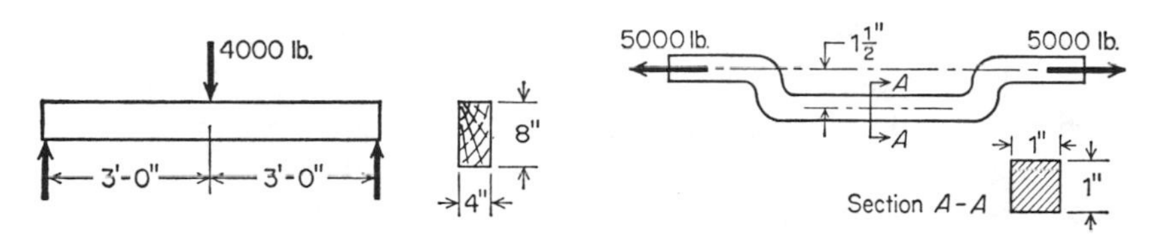

Figure 8.13 **Figure 8.14**

8.15. What is the maximum tensile stress at section A–A of the figure 8.14?

Ans. 50 ksi

8.16. (a) For the beam shown in figure 8.15, calculate reactions R_1 and R_2. (b) Draw the shear and moment diagrams. (c) Show the location and value of the maximum moment. (d) Show the location of and give the dimension to the point of inflection (where the moment changes from positive to negative). (e) What section modulus would be required if the allowable stress were 20,000 psi?

Ans. (a) 4290 lbf, 10,210 lbf, (c) 21,450 ft-lbf at $x = 5$ ft, (d) $x = 15.045$ ft, (e) 12.87 in^3

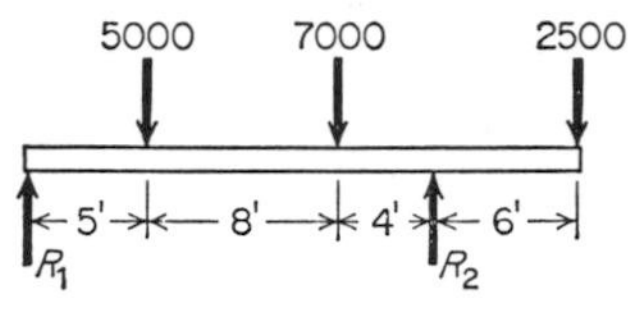

Figure 8.15

8.17. The length of a column divided by "r" (l/r ratio) is one of the criteria for the buckling of that column when subjected to compression loads. What does "r" stand for in this case? (a) The diameter of the column, (b) the least radius of gyration of the column, (c) the moment of inertia of the column, (d) the least dimension from the center to the outside of the column, (e) the restraining force.

Ans. (b)

8.18. The maximum bending moment in a loaded simple beam occurs: (a) at a section at which the shear is zero, (b) at the end supports, (c) at a section at which the shear is maximum, (d) at a section at which the maximum load occurs, (e) at none of these.

Ans. (a)

8.19. A steel punch measures 0.750 inch in diameter with no force applied to it. In punching a hole in a plate, a total compressive force of 35,000 lbf acts upon the punch. E is 3 EE7 psi and the Poisson ratio is 0.25 for this punch. What will be the actual diameter of the punch during the time it is so loaded?

Ans. 0.7505 inch

8.20. A steel roller is to be shrunk on a 2.000-inch diameter steel shaft by cooling the shaft with dry ice and heating the roller. Calculations show that the hole in the roller should be bored to give an interference of 0.003 inch at 70 °F to obtain the proper holding force. A clearance of 0.002 inch is desired when the roller and shaft are at their respective elevated and lowered temperature for ease in assembly. To what temperature must the roller be heated for assembly if the shaft can be cooled from 70 °F to 35 °F? Linear coefficient of expansion for steel is 6.5 EE–6 in/in-°F.

Ans. 420 °F

8.21. A wooden beam has an allowable bending stress of 1200 psi and is loaded as shown in figure 8.16. (a) Draw a shear diagram. (b) Draw a moment diagram. (c) Find the required section modulus for the beam. Neglect the weight of the beam.

Ans. (c) 150 in^3

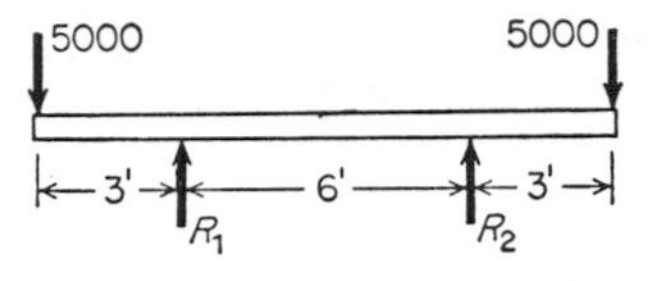

Figure 8.16

8.22. A 3-inch diameter solid steel shaft is used to transmit 24,000 in-lbf of torque at 99 rpm. (a) What is the maximum unit shearing stress in the shaft? (b) How much horsepower does the shaft transmit?

Ans. (a) 4530 psi, (b) 37.7

8.23. A steel rod containing a turnbuckle has its ends attached to rigid walls and is tightened by the turnbuckle in summer when the temperature is 90 °F to give a stress of 2000 psi. What will be the stress in the rod in winter when its temperature is 20 °F below zero?

Ans. 23,400 psi

8.24. The American Institute of Steel Construction recommends that the following formula be used for the design of steel columns:

$$P/A = 17{,}000 - 0.485(L/r)^2$$

P = total axial load in lbf, A = cross-section area in in^2, L = length of column in inches, and r = least radius of gyration in inches. What axial load will a steel column $1\frac{1}{2} \times 2 \times 36$ inches long support?

Ans. 40,980 lbf

8.25. A hollow shaft, with outside and inside diameters of 18 inches and 12 inches, respectively, transmits 10,000 hp at 99 rpm. What is the maximum unit shearing stress in the shaft?

Ans. 6920 psi

8.26. The distance between two points is exactly 100 feet. It is read correctly on a surveyor's tape when used at a temperature of 70 °F and subjected to a pull of 10 lbf. If the tape reads 100 feet when used at

100 °F under a pull of 20 lbf, what is the correct distance being measured? The tape is $\frac{1}{32}$ inch thick and $\frac{3}{8}$ inch wide.

Ans. 100.02234 ft

8.27. A cantilever beam 4 feet long has a concentrated load of 600 lbf at its end. Draw and label the shear and moment diagrams for this beam, and calculate the section modulus required for a maximum flexure stress of 1800 psi. Neglect the weight of the beam.

Ans. 16.0 in^3

8.28. A steel tie rod 1 inch $\times$ 2 inches $\times$ 12 feet long has each end fastened in a wall in such a manner that no slipping can take place. If the temperature drops from 110 °F to 20 °F, what tensile load is placed in the rod?

Ans. 35,100 lbf

8.29. Calculate the ratio of strength (torque-carrying capacity) for two steel shafts. The first is 2 inches in diameter and solid. The other is the same outside diameter but with a 1-inch hole axially through the middle. Assume that the maximum shear stress is the same for both shafts.

Ans. 16/15

8.30. A steel gear is to be assembled on a steel motor shaft by heating the gear and cooling the shaft with dry ice. At 70 °F, the shaft measures 7.000 inches in diameter and the hole in the gear measures 6.996 inches in diameter. To what temperature must the shaft be cooled so that the gear can be assembled on the shaft with 0.005 inch clearance when the gear is warmed to 140 °F? The coefficient of linear expansion for steel is 6.5 EE–6 in/in-°F.

Ans. −57.5 °F

8.31. A 6 $\times$ 2 $\times$ 10.5-lbf steel channel purlin (cross section area 3.07 in^2) runs 80 feet from wall to wall along the roof of a building. The ends were bolted to the walls when the temperature was 110 °F. When the temperature fell to 60 °F, the ends pulled loose from the concrete wall, due to improper design. What maximum force could have been exerted by the purlin on the wall?

Ans. 29,900 lbf

8.32. A contractor adds an excess amount of water to his concrete to make a more workable mix. Which of the following is true? (a) Excess water has little effect on the strength of the concrete. (b) As long as the aggregates are not allowed to settle, the concrete will not be affected by excess water. (c) Once the concrete is poured in the forms, it can be vibrated to mix up the aggregates, and the water that comes to the surface can be drained off, with little effect on the strength of the batch. (d) Excess water in concrete lowers the strength and care should be taken to keep the water-cement ratio to 7 gal/sack or less. (e) The only reason excess water in concrete is detrimental is because it causes the forms to warp and much concrete is lost through the cracks in the forms.

Ans. (d)

8.33. It is common practice for someone designing a reinforced concrete beam to assume: (a) that the concrete has one-third the tensile strength of the steel, (b) that the tensile strength of the concrete can be neglected, (c) that the modulus of elasticity of concrete and steel are the same, (d) that the steel has 2000 psi tension induced in it by the setting of the concrete, (e) that the weight of the beam can be neglected for long spans.

Ans. (b)

8.34. Sketch curves, with coordinates shown in figure 8.17, (a) showing the relation between the compressive strength of 28-day concrete and the water ratio in gallons per sack of cement, and (b) showing the relation between the compressive strength of 6 gal/sack concrete and the setting time. *Note*: Curves should be approximately correct in numbers and shape.

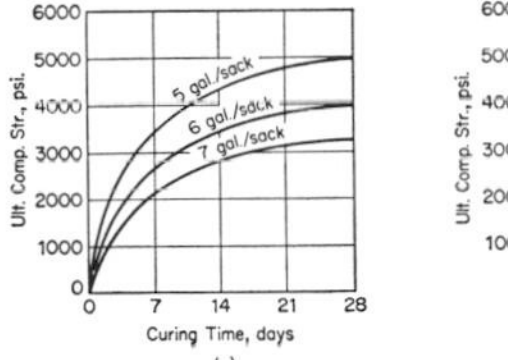

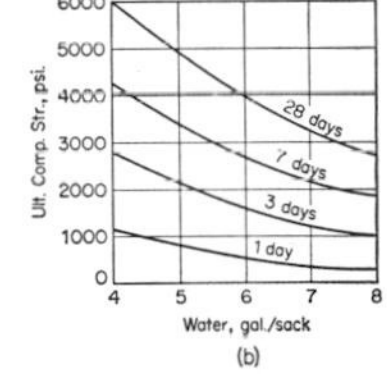

Figure 8.17

8.35. In the composite beam shown in figure 8.18, a wooden timber 8 inches wide by 12 inches deep has a steel plate securely fastened to its underside. The steel plate is $\frac{1}{4}$ inch thick and 8 inches wide. The moduli of elasticity are as follows: wood, 1.5 EE6 psi; steel 3 EE7 psi. Find y-bar, the distance from the upper surface to the neutral axis of the composite beam.

Ans. 7.8 inches

8.36. A simply-supported concrete beam with a value of $n = 10$ is subjected to a moment of 84,000 ft-lbf at the cross section shown in figure

8.19. What are the design stresses in concrete and steel? Each of the steel reinforcing bars is 1 inch square.

Ans. 1050 psi, 17,460 psi

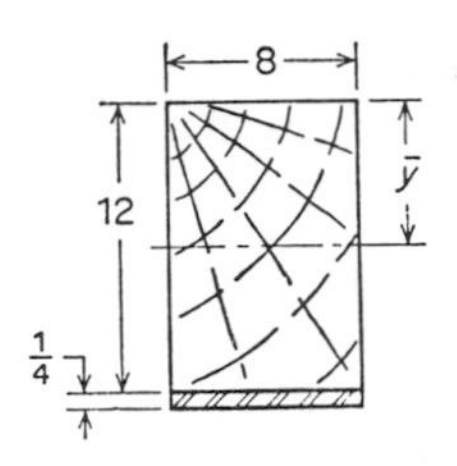

Figure 8.18

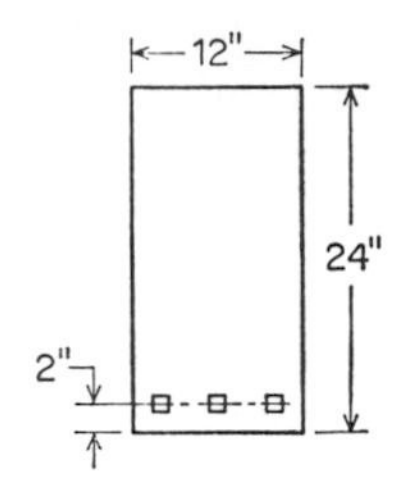

Figure 8.19

8.37. A timber cantilever beam 2 inches × 12 inches × 6 feet long is used to support a uniform load of 100 lbf/ft. E is 1.2 EE6 psi. What is the maximum deflection? The maximum allowable deflection is usually 1/360 of the span. Would this beam be satisfactory?

Ans. 0.081 inch, yes

8.38. A simple beam has a span of 12 feet supporting a concentrated load of 4000 lbf at the center of the span. The beam is a composite beam made of a rectangular wooden timber 4 inches wide by 6 inches deep and two steel plates, one above and one below the wood. The steel plates are $\frac{1}{8}$ inch thick and 4 inches wide. Calculate the maximum unit stress in the steel and in the timber. The moduli of elasticity of the timber and steel are 1 EE6 psi and 3 EE7 psi, respectively.

Ans. 37,200 psi, 1240 psi

8.39. A small machine part acts as a cantilever beam. What will be the ratio of deflection of two beams, one having a cross section $\frac{1}{4}$ inch × $\frac{1}{4}$ inch, the other having a cross section of $\frac{3}{8}$ × $\frac{3}{8}$ inch, all other dimensions and load being the same?

Ans. 5.06

8.40. Locate the neutral axis in figure 8.20 by calculating the distance y on the reinforced concrete beam section as shown. E of steel is 3 EE7 psi, E of concrete is 2 EE6 psi, and A_s area of steel is 3.00 in^2.

Ans. 7.28 inches

8.41. A machine part is loaded as a cantilever beam. Two possibilities (see figure 8.21) of design are being considered, utilizing different materials

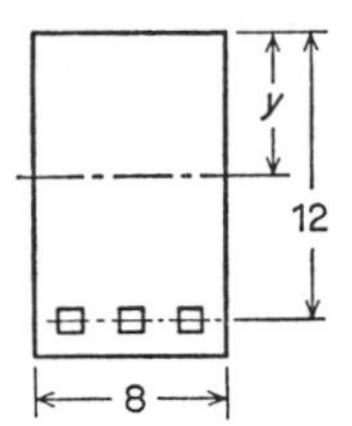

Figure 8.20

and different cross sections. Show by your calculations which design will be the more rigid.

Ans. number 1 by 17%

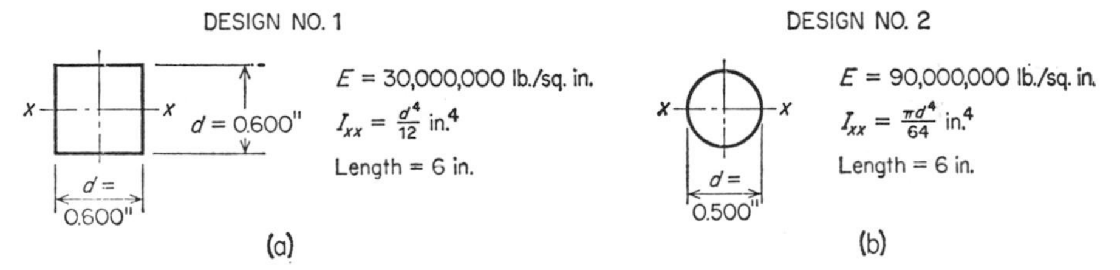

Figure 8.21

8.42. A 2 inch × 2 inch × 30 inch (actual dimensions) Douglas fir beam is tested to destruction in the laboratory. The beam was simply supported at each end and a concentrated load applied at the center of the span. Deflection is $PL^3/(48EI)$. The following tabulated data were obtained.

load (lbf)	deflection (inch)
100	0.025
200	0.050
300	0.080
500	0.130
...	...
915	failure occurred

(a) What was the average magnitude of the modulus of elasticity? (b) What was the magnitude of the *maximum shearing stress* at the cross section of the beam over each end support when failure occured? (c) What was the magnitude of the average shearing stress at the cross section directly under the load at a deflection of 0.050 inch? (d) At what bending stress did the beam fail?

Ans. (a) 1.65 EE6 psi, (b) 171.6 psi, (c) 0 psi, (d) 5150 psi

8.43. A weight of 3.22 lbm is attached to the end of a 0.0314 in^2 cross-sectional-area metal rod. The static stress-strain curve for this rod is shown in figure 8.22. The rod is then rotated about point O at a certain constant speed, which is maintained for a long period of time. The plane of rotation is horizontal and the original length of the rod was 19.000 inches. However, the length of the rod was found to have permanently increased by 0.038 inch as a result of the centrifugal force. What was the speed of rotation, assuming a negligible mass for the rod and considering only centrifugal forces?

Ans. 1150 rmp

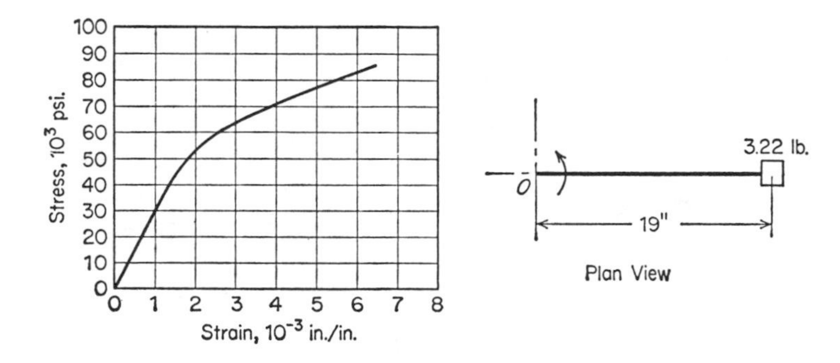

Figure 8.22

8.44. A cantilever steel beam is 12 feet long. The load is 6000 lbf/ft extending for 6 feet from support. Modulus of elasticity is 3 EE7 psi. The cross section of the beam is T-shaped, with a horizontal rectangle 6 inches wide by 2 inches thick and a vertical rectangle 12 inches deep by 1 inch thick. What is the maximum deflection of this beam in inches? Neglect the weight of the beam.

Ans. 0.295 inch

8.45. In the beam cross section shown in figure 8.23, the total vertical shear on the section is +40,000 lbf. The bending moment at the section is +120,000 ft-lbf. What are the maximum and minimum principal unit stresses at point A (at junction of the flange and web, but in the web)?

Ans. 16,375 (T), 2725 psi (C)

8.46. A cantilever beam of length L as one end built into a wall. The other end is simply supported. If the beam weighs w lbf/ft, its deflection y feet at a distance x feet from the built-in end is given by

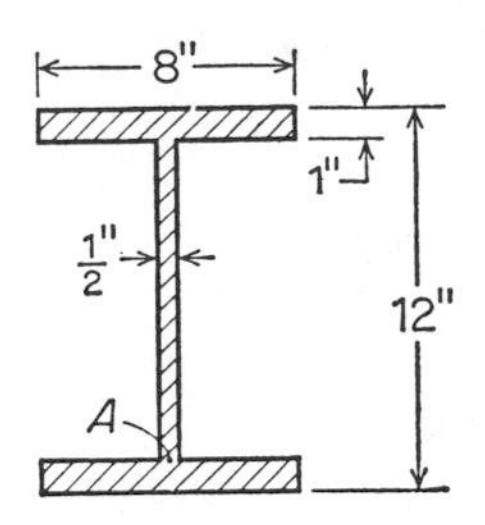

Figure 8.23

$$48EIy = w(2x^4 - 5Lx^3 + 3L^2x^2)$$

E and I are constant. How far from the built-in end does the maximum deflection occur?

Ans. $0.578L$ ft

8.47. A beam 25 feet long is simply supported at the left end and at 5 feet from the right end. It carries a uniformly distributed load of 2 kips/ft over a length of 10 feet starting at the left end of the beam. It also carries a concentrated load of 10 kips at the right end. Draw the shear and moment diagrams for this beam. Show all critical values and dimensions.

Ans. maximum $M = 50$ ft-kips

8.48. A short compression member, 8 inches high and 4 inches $\times$ 4 inches in cross section, is prepared by pouring copper around a 2 inch $\times$ 2 inch steel core. Calculate the unit stress in each material if the member supports an axial load of 50 kips. $E_{st} = 30{,}000$ ksi and $E_{cu} = 15{,}000$ ksi.

Ans. Cu: 2500 psi, steel: 5000 psi

8.49. A simply-supported T-beam carries a concentrated load at the middle of a 10-foot span. The cross section is made up with two full-size 2-inch $\times$ 10-inch planks. The plank in the vertical position is placed directly under the center of the horizontal plank and securely spiked. Neglecting the weight of the beam itself, compute the maximum allowable concentrated load. Allowable stresses: 1200 psi in bending, 120 psi in shear.

Ans. 2667 lbf

8.50. Draw shear and moment diagrams, locating all critical points, for a simply supported beam 24 feet long if the beam is supported at the right end and at 4 feet from the left end. It carries a uniformly distributed load of 2 kips/ft between supports and a concentrated load of 6 kips at the left end.

Ans. maximum $M = 88.4$ ft-kips

8.51. A column is to be inserted next to an existing column to take up some of the load. The new column is to be 10 feet, stressed to 10,000 psi. The column will be shortened sufficiently to be inserted in place by cooling with dry ice. Assuming a normal temperature of 70 °F, to what temperature should the column be cooled to provide a clearance of $\frac{1}{32}$ inch before insertion and yet take up its load when it reaches room temperature? The modulus of elasticity is 3 EE7 psi and the temperature coefficient is 6.7 EE–6.

Ans. −18.3 °F

8.52. Where does the maximum bending moment occur in a beam with a 24-foot span that carries a load of 24,000 lbf uniformly spread over its entire length and a further load of 12 tons uniformly spread over a section that starts 6 feet from the left support and extends 8 feet to the right? What is the maximum moment?

Ans. 11 ft from left end, 188 ft-kips

8.53. A simply-supported timber beam is 2 inches wide, 8 inches deep, and 12 feet long. It supports a concentrated load of 1000 lbf at a point 4 feet from the left end. Neglect the weight of the beam. Modulus of elasticity is $E = 1.5$ EE6 psi. Find the magnitude and location of the maximum deflection.

Ans. 0.418 inch, 5.47 ft from left end

8.54. A timber 4 inches × 4 inches cross section and 24 feet long is used as a column. The modulus of elasticity is 1.5 EE6 psi. The ends are free to turn or rotate, but may not be displaced. An average unit stress of 1500 psi is allowable. Find the maximum load that this column can support.

Ans. 3820 lbf (Euler)

8.55. A steel bar 6 inches wide and $\frac{1}{2}$ inch thick has a copper bar 6 inches wide and $\frac{1}{8}$ inch thick on each side so that the total cross section of the combined bar is 6 inches by $\frac{3}{4}$ inch. A tensile load of 80 kips is applied to the ends of the bar in a direction perpendicular to its cross section. Find the unit stress and the unit strain caused by this axial load in each of the two materials. $E_s = 3$ EE7 psi, $E_{\text{cu}} = 1.5$ EE7 psi.

Ans. steel: 21,333 psi; copper, 10,667 psi; each 7.1 EE–4 inch/inch

8.56. Design a simple beam for the following conditions: ratio of width/depth = 1/2; material, wood; maximum fiber stress = 1200 psi; length of beam = 20 ft; load = 6000 lbf concentrated 6 ft from the right end. $I = bd^3/12$.

Ans. 7.21 inches × 14.42 inches

8.57. A 12-inch, 40.8-lbm I-beam is used as a building column. Its unsupported length is 14 feet. What safe axial load will it carry? Use a recognized column formula and name reference (title and page). Show all calculations and formulas.

Ans. 91 kips, (NY formula)

8.58. A reinforced concrete beam is on a simple span of 6 feet and carries a concentrated load of 10 kips at each one-third point. The beam section is 10 inches wide by 18 inches deep with an effective depth of 16 inches, and is reinforced with two $\frac{7}{8}$-inch round bars. If $n = 15$, what are the maximum stresses in the longitudinal steel and in the concrete?

Ans. 14.25 ksi, 572 psi

8.59. A steel rod $\frac{1}{2}$ inch square in cross section and 20 feet long is suspended from a ceiling. What load hung from its lower end will stretch it 0.2 inch?

Ans. 6242 lbf

8.60. A simply-supported timber 6 inches × 6 inches in cross section and 12 feet long will safely carry a uniformly distributed load of 2400 lbf. If used as a cantilever beam with a concentrated load as its free end, it would safely carry what load?

Ans. 300 lbf

8.61. A plank 4 inches wide, 12 inches deep, and 12 feet long used as a simple beam with a concentrated load of 4000 lbf applied at mid-span will deflect 0.3 inch when the 12-inch side is vertical. How much will it deflect under the same load if it is placed with the 4-inch side vertical?

Ans. 2.7 inches

8.62. A timber cantilever beam 10 inches wide by 2 inches deep and 10 feet long has a load of 150 lbf at a point 3 feet from the free end. Find the slope and deflection at the free end if $E = 1.5$ EE6 psi. Determine the maximum bending stress, also.

Ans. 0.053, 4.87 inches, 1890 psi

8.63. A square steel bar of 1 inch × 1 inch cross section and 6 feet long is to be used as a column. The modulus of elasticity is $E = 3$ EE7 psi. The ends are free to rotate but may not be displaced. An average unit stress of 30 ksi is allowable. Find the maximum load that this column can support.

Ans. 4750 lbf (Euler)

8.64. A beam 40 feet long rests on simple supports at the ends and has the following loads: dead load = 300 lbf/ft; uniform live load = 700 lbf/ft; concentrated central live load = 10,000 lbf. At a point 16 feet from the left support, find (a) the maximum and minimum bending moments that can be developed, (b) the maximum and minimum shear.

Ans. (a) $M_{max} = 272$ ft-kips, (b) $V_{max} = 9$ kips

8.65. A steel shaft, 4 inches in diameter, transmits 200 hp when rotating at a speed of 240 rpm. Find the maximum shearing unit stress in the outer fiber. Use $E = 3$ EE7 psi.

Ans. 4170 psi

8.66. A timber beam with a cross section 8 inches wide by 12 inches deep is 16 feet long. It is supported as a simple beam with one support at the left end and the other 4 feet from the right end. There is a concentrated load of 4800 lbf on the overhanging end and another concentrated load of 2000 lbf placed 3 feet from the left end. The span between supports also has a uniform load of 600 lbf/ft. Plot the shear diagram and find the maximum unit tensile stresses due to the positive and negative moments. Also check for unit shearing stress. Is the beam safe?

Ans. 487 psi, 1200 psi, 89 psi; yes

8.67. A steel wire accurately graduated in feet was used to measure the depth of a deep well. The cross-sectional area of the wire was 0.03 in^2 and a 250-lbf weight was attached to the wire to lower it into the well. According to the graduated scale, the measurement read 2000 feet. What should the corrected measurement be? A steel bar of 1 in^2 cross-sectional area and 1 yard long weighs 10 lbf.

Ans. 2000.778 ft

8.68. A column is built up of two 12-inch, 35-lbf channels laced together so that the back-to-back distance of the channels is such that the moments of inertia about the two principal axes are equal with the flanges of the channels pointing outward. The column is 20 feet long. Calculate the working load. The eccentricity is zero and the ends are considered fixed.

Ans. 341 kips (AISC)

8.69. A cantilever beam 8 feet long is fixed at the left end. It carries a uniformly distributed load of 100 lbf/linear ft, including its own weight, and a concentrated load of 1000 lbf at the free end. Write the bending moment equation for any point along the beam.

Ans. $M = -11{,}200 + 1800x - 50x^2$ ft-lbf

8.70. An 8-inch, 40-lbf wide-flange beam (8×8 WF40) used as a column is 30 feet long. It is supported at the middle in the direction normal to the web but is unsupported in the direction parallel to the web. The area of the section is 11.76 in^2, radius of gyration about axis perpendicular to web is 3.53 inches, and radius of gyration about axis parallel to web is 2.04 inches. The factor of safety is 3. E = 3 EE7 psi. Assume round end conditions. By means of Euler's equation, find the safe load on the column.

Ans. 111.7 kips

8.71. Compute the size of a circular steel shaft to transmit 100 hp at 1000 rpm with an allowable extreme fiber stress in shear of 10,000 psi. The angle of twist must not be greater than 1° per foot of length of shaft. The modulus of elasticity in shear is 1.2 EE7 psi.

Ans. 1.475 inches

9 *THERMODYNAMICS*

SYMBOLS

A = area
C = a constant
c = specific heat; c_p at constant pressure, c_v at constant volume
E = general symbol for energy; total stored energy
e = thermal efficiency
g = acceleration of gravity, (32.2) ft/sec^2
g_c = standard acceleration of gravity, lbm-ft/lbf-sec^2
H = total enthalpy
h = specific enthalpy, BTU/lbm
hp = horsepower
J = Joule's constant (≈ 778) ft-lbf/BTU
K = kinetic energy
k = c_p/c_v
M = molecular weight (lbm/mole)
m = mass in lbm
n = rpm; polytropic exponent
P = potential energy of gravitation
p = unit pressure (psf)
Q = heat; *net* heat in energy equations
R = specific gas constant; $R = \overline{R}/M = 1545/M$

$\overline{R}$ = universal gas constant ($\approx$ 1545) ft-lbf/lbm-mole-°R
S = total entropy
s = specific entropy (BTU/lbm-°R)
T = absolute temperature; on Fahrenheit scale, called degrees Rankine, °R; on Celsius scale, called degrees Kelvin, °K
U = total internal (molecular) energy
u = specific internal energy (BTU/lbm)
V = total volume
v = specific volume, cu ft/lbm
$\mathcal{V}$ = velocity, speed (script vee to distinguish from specific volume)
W = work (the kind that is transmitted by a shaft); *net* work in energy equations
x = quality of a two-phase system
y = fraction of liquid in two-phase system
z = altitude; approximate potential energy of 1 lbm (ft-lbf/lbm)
η = efficiency ratios; engine efficiency
ρ = density (lbm/cu ft)
ϕ = relative humidity
ω = humidity ratio, lbm vapor/lbm dry gas
Δ indicates a difference or change

FUNDAMENTAL CONSIDERATIONS

Definitions

There are too many definitions to give all of them here. For the most part, definitions must be abbreviated to equation forms.

$$\textit{density} = \frac{\text{mass}}{\text{unit volume}} = \frac{1}{v} = \rho \text{ lbm/cu ft} = \frac{\rho}{g_c} \text{ slugs/cu ft}$$

$$\textit{specific heat} = \frac{\text{heat (energy units)}}{\text{(unit mass)(change of temperature)}}$$

$$c_p = \left(\frac{dh}{dT}\right)_p = \frac{Q]_p}{m\Delta T} = \frac{h_2 - h_1}{\Delta T} \text{ BTU/lbm-°R}$$

$$c_v = \left(\frac{du}{dT}\right)_v = \frac{Q]_v}{m\Delta T} = \frac{u_2 - u_1}{\Delta T} \text{ BTU/lbm-°R}$$

[MOST IDEAL GAS PROBLEMS]

Subscripts p and v mean that these properties are held constant.

entropy, $\Delta S = \int dQ/T$
$dQ = TdS$ [REVERSIBLE PROCESS]

enthalpy, $h = u + pv/J$ BTU/lbm

gravitational potential energy (m = mass in lbm), $P = mgz/g_c \approx mz$ ft-lbf

kinetic energy, $K = mv^2/2g_c$ ft-lbm

flow work (flow energy), $W_f = pV/J$ BTU
$= pv/J$ BTU/lbm
(This is appropriate only when accounting for this energy crossing a boundary into or out of a system.)

internal energy, U, u, stored within the system as (largely) molecular energy

Work, W, is determined in several ways, depending upon the kind of process or cycle and the working substance.

Heat, Q, is that which transfers from one body to another by virtue of a temperature difference. (This applies to an electromagnetic emanation—*never* anything *in* a body or system. Energy contained in a body or system is given the general name *stored energy*—sometimes *internal energy* in a broader sense than molecular to include kinetic energy, potential energy, and other forms that can be stored.)

There are a number of other forms of energy, but in the following discussions, it is assumed that such other forms, including chemical energy, electricity, radio waves, surface tension, and others, are irrelevant, unless specifically included.

Laws of Thermodynamics

Zeroth law: If two bodies are in thermal equilibrium with a third body, the two bodies are in thermal equilibrium with each other.

First law: Energy can be neither created nor destroyed, and one form of energy can be transformed into another form.

Second law: Net heat will not flow of its own accord from a cold body to a hot body (Clausius). It is impossible to construct an engine that, while operating in a cycle, produces no effect except to do work and exchange heat with a single reservoir (Kelvin-Planck). Work can be converted entirely into heat, but heat cannot be converted entirely into work continuously for an indefinite period (even theoretically).

Working Form of First Law

One of two points of view is useful:

$$\begin{bmatrix}\text{energy}\\ \text{entering}\\ \text{system}\end{bmatrix} = \begin{bmatrix}\text{increase (or decrease—negative)}\\ \text{of stored energy within the system}\\ (U \text{ for nonflow, } E \text{ for general case})\end{bmatrix} + \begin{bmatrix}\text{energy}\\ \text{leaving}\\ \text{system}\end{bmatrix}$$

$$\begin{bmatrix}\text{initial}\\ \text{stored}\\ \text{energy}\end{bmatrix} + \begin{bmatrix}\text{energy}\\ \text{entering}\\ \text{system}\end{bmatrix} - \begin{bmatrix}\text{energy}\\ \text{leaving}\\ \text{system}\end{bmatrix} = \begin{bmatrix}\text{final}\\ \text{stored}\\ \text{energy}\end{bmatrix}$$

For any closed system,

$$Q = E_2 - E_1 + W$$
$$dQ = dE + dW$$

(E is total stored energy; Q is *net* heat flow; W is *net* work done.)

For any nonflow closed system,

$$Q = U_2 - U_1 + W$$

Or,

(a)
$$dQ = dU + dW \quad \text{[FOR } m \text{ POUNDS]}$$
$$dQ = du + dW \quad \text{[FOR 1 POUND]}$$

For any steady-flow open system (no change in stored energy),

(b)
$$Q = U_2 - U_1 + \frac{p_2V_2}{J} - \frac{p_1V_1}{J} + K_2 - K_1 + P_2 - P_1 + W$$

$\Delta P = P_2 - P_1$ is ordinarily negligible in a thermal system. Drop ΔP, and use $H = U + pV/J$:

(c)
$$Q = H_2 - H_1 + K_2 - K_1 + W = \Delta H + \Delta K + W$$

Or,

$$dQ = dH + dK + dW \quad \text{[FOR } m \text{ POUNDS]}$$
$$dQ = dh + dK + dW \quad \text{[FOR 1 POUND]}$$

For these energy equations, increases of stored forms of energy are positive, and decreases are negative. Heat *added* to a system is positive, and heat *rejected* is negative. Work done *by* a system is positive, and work done *on* a system is negative.

Work

If all the other terms are known in one of the foregoing energy equations, W can be computed from it.

For a nonflow system (reversible processes):

$$W = \int p dV \text{ ft-lbf}$$

For a steady-flow system (reversible processes):

(d) $$\int p\,dV = \Delta W_f + \Delta K + \Delta P + W \text{ ft-lbf}$$

(e) $$-\int V\,dp = W + \Delta K \text{ ft-lbf} \qquad [\Delta P = 0]$$

Work is also obtained from indicator cards, called *indicated work*, and from various test devices that measure the work (and power) being transmitted along a shaft, called *shaft work* or *brake work*.

Heat

Heat is variously computed from one of the energy equations, or from

(f) $$Q = m\int c\,dT = mc(T_2 - T_1) \qquad \text{[REVERSIBLE]}$$

[CONSTANT C]

(g) $$Q = \int T\,dS$$

$$\Delta S = \int \frac{dQ}{T} \qquad \text{[REVERSIBLE]}$$

c depends on the kind of process and temperature as well as on the substance. Mathematical relations for any reversible processes are:

(h) $$T\,ds = du + \frac{p\,dv}{J}$$

$$dQ = du + \frac{p\,dv}{J}$$

(i) $$T\,ds = dh - \frac{v\,dp}{J}$$

$$dQ = dh - \frac{v\,dp}{J}$$

These equations are also valid for any substance, but the integrations are usually not easily made except for an ideal gas.

EXAMPLE: A 50-lbm piece of iron at 200 °F is submerged in 1 cu ft of water at 70 °F. For iron, $c = 0.1$ BTU/lbm-°F. Assume there is no heat loss from the system. Compute the final temperature.

SOLUTION: You are supposed to know that for water, $c = 1$ and that the mass of 1 cu ft of water is approximately 62.4 lbm/cu ft. By the law of conservation of energy, energy given up by iron is equal to energy received by water (or, the total stored energy of the system remains constant). Let $T_2 =$ final temperature of water and iron at thermal equilibrium.

$$(50)(0.1)(200 - T_2) = (62.4)(1)(T_2 - 70)$$

From this,

$$T_2 = 79.7\ °\mathrm{F}$$

Reversibility

If, after a process consisting of a continuous series of equilibrium states is completed, the substance can be made to retrace in the reverse order the various states of the original process, and if all energy quantities to or from the surroundings can be returned to their original states, then the process is externally and internally reversible. If a process is *internally* irreversible, due to fluid or other friction for instance, the thermodynamic expressions in integral form, as $\int p\,dV$, lose their previously defined meanings.

IDEAL GAS

An ideal gas is also called a *perfect gas.*

Laws of Ideal Gas

Boyle: When T is constant,

$$pV = C$$
$$p_1V_1 = p_2V_2$$

Charles: When p is constant,

$$\frac{T}{V} = C$$
$$\frac{T_1}{V_1} = \frac{T_2}{V_2}$$

When V is constant,

$$\frac{T}{p} = C$$

$$\frac{T_1}{p_1} = \frac{T_2}{p_2}$$

(j) Equation of state:

$$pV = mRT$$

$$\frac{P_1 V_1}{T_1} = \frac{P_2 V_2}{T_2}$$

Also,

$$pv = RT$$

$$p = \rho RT$$

$$H = U + \frac{pV}{J} = U + \frac{mRT}{J}$$

$$h = u + \frac{RT}{J}$$

R is in ft-lbf/lbm-°R.

For air:

$$R = 53.3$$

$$c_p = 0.24 \text{ BTU/lbm-°R}$$

$$c_v = 0.1715 \text{ BTU/lbm-°R}$$

$$k = \frac{c_p}{c_v} = 1.4$$

[NORMAL TEMPERATURES]

(k) Joule's law:

$$dU = mc_v \, dT$$

$$\Delta U = m \int c_v \, dT = mc_v(T_2 - T_1) \qquad \text{[CONSTANT } c_v\text{]}$$

(l)

$$dH = mc_p \, dT$$

$$\Delta H = m \int c_p \, dT = mc_p(T_2 - T_1) \qquad \text{[CONSTANT } c_p\text{]}$$

(m)

$$c_p - c_v = \frac{R}{J}$$

[difference in specific heats is equal to gas constant in BTU units]

Entropy changes for any process, reversible or irreversible, of an ideal gas can be obtained by using equations (j), (k), and (l) in (h) and (i) and integrating.

Archimedes' Principle

The law of physics regarding buoyancy is often useful: *The force of buoyancy is equal to the weight of the displaced fluid*, and the line of action of the resultant vector passes through the center of gravity of the displaced fluid.

EXAMPLE: An observation balloon uses helium gas for buoyancy. If the balloon, observers, and equipment weigh 600 lbf, what must be the minimum volume of the balloon to realize buoyancy? The temperature and pressure of the atmosphere are 80 °F and 13 psia. The specific volumes of air and helium at 14.7 psia and 32 °F are 12.4 cu ft/lbm and 89.1 cu ft/lbm.

SOLUTION:

For air:

$$\frac{p_o v_o}{T_o} = \frac{p_a v_a}{T_a} = \frac{(14.7)(144)(12.4)}{32 + 460} = \frac{(13)(144)v_a}{80 + 460}$$

From this, the specific volume of air is $v_a = 15.4$ cu ft/lbm.

For helium:

$$\frac{(14.7)(144)(89.1)}{32 + 460} = \frac{(13)(144)v_h}{80 + 460}$$

Or,

$$v_h = 110.5 \text{ cu ft/lbm}$$

For balance, buoyancy − weight of He − load = 0.

Let total volume $= V$. Weight $= V/v$ (closely).

$$\frac{V}{15.4} - \frac{V}{110.5} - 600 = 0$$

$$V = 10{,}710 \text{ cu ft}$$

Note: If a table of values of R had been permitted, there would be superfluous data because $v = 1/\rho$ is obtained from $pv = RT$.

PROCESSES

All of the following equations come from the foregoing energy relations (and ideal gas laws when appropriate).

nonflow: $dW = dQ - dU$

steady flow: $dW = dQ - dH - dK$

Constant Volume Process (Reversible)

$$\int p\,dV = 0$$

$$Q = \Delta U = mc_v(T_2 - T_1)$$

[ANY SUBSTANCE] [IDEAL GAS]

ideal gas:

$$\Delta S = mc_v \ln\frac{T_2}{T_1}$$ [FROM (g)]

Constant Pressure Process (Reversible)

$$Q = \Delta H \quad = mc_p(T_2 - T_1)$$

[ANY SUBSTANCE] [IDEAL GAS]

nonflow:

$$W = \int p\,dV = p(V_2 - V_1) \text{ ft-lbf}$$ [ANY SUBSTANCE]

ideal gas:

$$Q = \Delta U + \frac{p\,dV}{J}$$

$$mc_p dT = mc_v dT + \frac{p\,dV}{J}$$

$$\Delta S = mc_p \ln\frac{T_2}{T_1}$$ [FROM (g)]

Isothermal Process ($T = C$, Reversible)

$$\Delta s = \int \frac{dQ}{T} = \frac{Q}{T}$$

$$Q = T(S_2 - S_1)$$

$$\Delta u = \left(h_2 - \frac{p_2 v_2}{J}\right) - \left(h_1 - \frac{p_1 v_1}{J}\right)$$

nonflow: $W = Q - \Delta u = \int p\,dV/J$

[ANY SUBSTANCE]

steady flow: $W = Q - \Delta h - \Delta K$

ideal gas:

$$pV = C$$
$$T = C$$

Therefore,

$$\Delta u = 0$$
$$\Delta h = 0$$
$$u = C$$
$$h = C$$
$$\int p\,dv = -\int v\,dp = p_1 v_1 \ln\frac{v_2}{v_1} \text{ ft-lbf/lbm}$$
$$Q = \int \frac{p\,dv}{J} \qquad \text{[FROM (h)]}$$
$$Q = -\int \frac{v\,dp}{J} \qquad \text{[FROM (i)]}$$

steady flow, from (e):

$$W = -\int \frac{v\,dp}{J} + \Delta K \qquad [\Delta P = 0]$$

Adiabatic Process

By definition, $Q = 0$ (reversible or irreversible).

nonflow: $W = -\Delta u = -(u_2 - u_1)$

[ANY SUBSTANCE]

steady flow: $W = -(\Delta h + \Delta K)$

Reversible adiabatic process = *isentropic process*,

$$s_1 = s_2$$
$$\Delta s = 0 \qquad \text{[ANY SUBSTANCE]}$$

ideal gas, isentropic process:

$$pV^k = C$$

(n)
$$\int p\,dV = \frac{p_2V_2 - p_1V_1}{1-k}$$
$$-\int V\,dp = \frac{k(p_2V_2 - p_1V_1)}{1-k} \text{ ft-lbf}$$

nonflow:
$$W = \int p\,dV/J = -\Delta U = -mc_v(T_2 - T_1)$$

steady flow:
$$W = -(\Delta H + \Delta K) = -mc_p(T_2 - T_1) \qquad [\Delta K = 0]$$
$$W = -\int \frac{V\,dp}{J} = -\frac{k(p_2V_2 - p_1V_1)}{J(1-k)}\text{BTU} \qquad [\Delta K = 0]$$

(o)
$$\frac{T_2}{T_1} = \left(\frac{p_2}{p_1}\right)^{(k-1)/k} = \left(\frac{V_1}{V_2}\right)^{k-1}$$

Watch for the use of the word *adiabatic* instead of *isentropic*. Strictly, an adiabatic can be reversible or irreversible. If it is not specified, reversible is probably intended.

Flow through a nozzle is adiabatic, so $Q = 0$. No work is done, so $W = 0$. If the initial kinetic energy is negligible, the steady-flow energy equation reduces to
$$-\Delta h = \Delta K = \mathcal{V}_2^2/(2g_cJ) = \mathcal{V}_2^2/50{,}000$$

(p)
$$\mathcal{V}_2 = \sqrt{2g_cJ(h_1 - h_2)} = 223.8\sqrt{h_1 - h_2} = 223.8\sqrt{c_p(T_1 - T_2)}$$
[ANY EXPANSIBLE FLUID] [IDEAL GAS]

The continuity of mass equation also applies:

(q)
$$\frac{A_1\mathcal{V}_1}{v_1} = \frac{A_2\mathcal{V}_2}{v_2}$$
$$\rho_1A_1\mathcal{V}_1 = \rho_2A_2\mathcal{V}_2$$

If h_o = stagnation enthalpy = $h + K$ is used instead of h_1 above, there is an automatic correction for initial velocity $\mathcal{V}_1$. If the flow is frictionless (isentropic), $s_1 = s_2$, which is the key to the solution of problems with expansible fluids.

Polytropic Process (Reversible)

By definition, $pV^n = C$ and $p_1V_1^n = p_2V_2^n$.

$$-\int V\,dp = \frac{n(p_2V_2 - p_1V_1)}{1-n} \qquad \text{[ANY SUBSTANCE]}$$

$$\int p\,dV = \frac{p_2V_2 - p_1V_1}{1-n} \qquad \text{[ANY SUBSTANCE]}$$

$$Q = \Delta u + \int \frac{p\,dV}{J}\ \text{BTU} \qquad \text{[FROM (h)]}$$

$$Q = \Delta h - \int \frac{V\,dp}{J}\ \text{BTU} \qquad \text{[FROM (i)]}$$

All of the expressions under isentropic process with k in them apply for ideal gas, with $k = n$.

nonflow: $$W = \int p\,dV \text{ ft-lbf}$$

steady flow: $$W = -\Delta K - \int V\,dp \text{ ft-lbf} \qquad \text{[FROM (e)]}$$

any:

$$\Delta U = mc_v\Delta T \text{ BTU}$$

$$\Delta S = mc_n \ln\frac{T_2}{T_1} \text{ BTU/lbm-°R}$$

(r) $$Q = \Delta U + \int \frac{p\,dV}{J} = mc_n(T_2 - T_1) \text{ BTU}$$

$$c_n = c_v\left(\frac{k-n}{1-n}\right) \text{ BTU/lbm-°R}$$

Throttling Process (Irreversible Steady Flow)

$H_1 = H_2$ is used to compute the quality of wet stream.

Note: If the examination has a problem that says the substance undergoes a process, without saying whether it is nonflow or steady flow, nonflow is probably intended—unless the implication otherwise is clear, as in saying the process occurs in a turbine, which is analyzed as a steady flow machine.

EXAMPLE 1: 50 cu ft of air at 14.7 psia and 70 °F are compressed to 100 psia in a reversible process such that $pV^{1.3} = C$. Calculate (a) work of compression (ft-lbf), (b) change in internal energy (BTU), (c) heat added or rejected (BTU), (d) change of enthalpy (BTU), (e) change of entropy. State the directions in which these changes occur.

SOLUTION: (a) It does not say whether the process of compression is steady flow or non-flow. This fact affects only part (a), and one could solve for both works without much extra effort. For steady flow, assume $\Delta K = 0$. For a reversible nonflow process, $W = \int p\,dV$. The quickest solution is to find

$$T_2 = T_1 \left(\frac{p_2}{p_1}\right)^{(n-1)/n}$$

$$m = \frac{p_1 V_1}{R T_1}$$

Then,

$$W_{\text{non}} = \frac{p_2 V_2 - p_1 V_1}{1-n} = \frac{mR}{1-n}(T_2 - T_1) \text{ ft-lbf}$$

For steady flow, $W = -\int V\,dp$ [equation (e)], and for $pV^n = C$, this integral is $n \int p\,dV$, or

$$W_{\text{S.F.}} = \frac{nmR}{1-n}(T_2 - T_1) \text{ ft-lbf}$$

Work of steady flow includes the net work of intake and discharge.

(b) ideal gas: $\Delta U = mc_v(T_2 - T_1)$ BTU (increase)

(c) $Q = mc_n(T_2 - T_1)$ BTU, $c_n = c_v \dfrac{k-n}{1-n}$ (rejected)

(d) ideal gas: $\Delta H = mc_p(T_2 - T_1)$ BTU (increase)

(e) $\Delta S = mc_n \ln(T_2/T_1)$ BTU/°R (decrease)

EXAMPLE 2: A steam turbine is operating on a flow of 2 lbm/second of steam. The enthalpies are 1292 BTU/lbm and 1098 BTU/lbm entering and leaving, respectively. The entering and leaving steam velocities are 50 ft/sec and 117 ft/sec, respectively. There is a heat loss of 13 BTU/lbm of steam as it flows through the machine. Find the work done. The kinetic energy change is what percentage of the enthalpy change? Does the change of kinetic energy result in an increase or decrease of work?

SOLUTION: The system is the steam in the turbine between entrance and exit sections. The volume of such a system is sometimes called the *control volume.* See figure 9.1. For 1 lbm,

$$K_1 = \frac{\mathcal{V}^2}{2gJ} = \frac{(50)^2}{50{,}000} = 0.05 \text{ BTU/lbm}$$

$$K_2 = \frac{(117)^2}{50{,}000} = 0.274 \text{ BTU/lbm}$$

Using the steady-flow equation,

$$W = -\Delta h - \Delta K + Q = -(1098 - 1292) - (0.274 - 0.05) + (-13)$$
$$W = 180.78 \text{ BTU/lbm}$$

For 2 lbm/second, $W = 361.55$ BTU/second. (*Ans.*)

$$\frac{\Delta K}{\Delta h} = \frac{0.224}{194} = 0.001154 \approx 0.115\% \quad (\textit{Ans.})$$

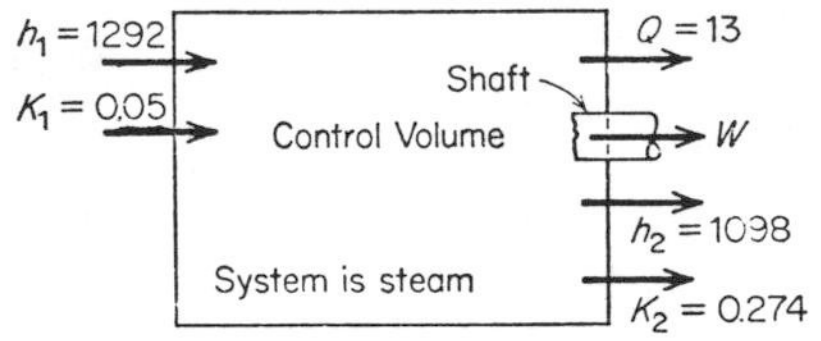

Figure 9.1 Energy diagram for turbine

In this case, work is slightly less than it would have been for $\Delta K = 0$; therefore, decrease. (*Ans.*)

Note: As in drawing free-bodies in mechanics, it is a safeguard to draw energy diagrams—a picture of the system with the energy flows represented thereon—as in figure 9.1. For no change in stored energy, energy entering is equal to energy departing, from which $W = 180.78$ BTU/lbm, as before.

CYCLES

In a cycle that is operating steady state (no change in stored energy or mass, and so forth),

$$\text{heat in} - \text{heat out} = \text{work done by} - \text{work done on}$$

$$\sum Q = W_{\text{net}}$$

This is valid for any substance, reversible or irreversible processes. A cycle always closes on the pV and TS planes (or any other)—it always returns to the same state at the same point in the cycle. Areas on the pV plane represent work. The net work is the algebraic sum of the integrals of $p\,dV$ for each reversible process of the cycle:

(s) $$W_{\text{net}} = \sum \int p\,dV \text{ ft-lbf}$$ [REVERSIBLE]

Areas on the TS plane represent heat ($dQ = T\,dS$). If all Q's are algebraically summed ($\sum Q$) for all processes of the cycle, the net work is obtained.

(t) $$\sum Q = \sum \int wc\,dT$$ [REVERSIBLE]

$$\sum Q = \sum \int T\,dS$$ [REVERSIBLE]

To evaluate by computation, the processes need to be internally reversible. Repeat *process* analysis until all sums are made.

Carnot Principle

The Carnot cycle is the most efficient cycle conceivable; it is reversible externally (as ΔT approaches zero, as in figure 9.2) and internally.

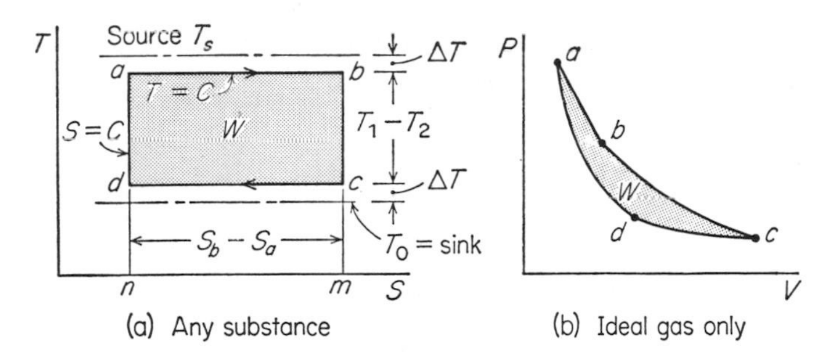

Figure 9.2 Carnot cycle

(u) $$\text{ideal thermal efficiency } e = \frac{W_{\text{net}}}{Q_A}$$ [ANY CYCLE]

$$\text{actual thermal efficiency } e' = \frac{W'_{\text{net}}}{E_c}$$

E_c is the energy chargeable against the actual cycle or engine.

$$\text{Carnot efficiency } e = \frac{T_1 - T_2}{T_1}$$

This is a thermal efficiency. See figure 9.2. The heat added in a Carnot cycle is represented by area $abmn = T_1(\Delta S)$. The heat rejected is represented by $cmnd = T_2\Delta S$. The work is represented by $abcd = (T_1 - T_2)\Delta S$.

Unavailable Energy

If an adiabatic system does work by an irreversible process, the increase in unavailable energy is

$$E_u = T_o\Delta S \quad \text{(BTU)}$$

T_o is the sink temperature and ΔS is the system's change of entropy during the process.

EFFICIENCIES

Engine Efficiencies

$$\text{(v)} \qquad \text{engine efficiency } \eta = \frac{\text{actual work}}{\text{corresponding ideal work}}$$

A reciprocating engine has an indicated engine efficiency η_i. All engines have brake or shaft engine efficiencies η_b. An engine that drives a generator has an over-all or combined engine efficiency η_k. Correspondingly, there are three places to measure the power or work.

$$\eta_i = \frac{W_i}{W}$$

$$\eta_b = \frac{W_b}{W}$$

$$\eta_k = \frac{W_k}{W}$$

W_i = indicated work, W_b = brake work, W_k = combined work, and W = work of the corresponding ideal engine. If the engine efficiency is given without a qualifying adjective, the kind involved must be decided by the context. It probably is brake (or shaft) engine efficiency. Generally,

$$\text{mechanical efficiency } \eta_m = \frac{\text{output}}{\text{input}}$$

This is the same as for a generator. Applied to a reciprocating engine,

$$\eta_m = \frac{W_b}{W_i} = \frac{\text{bhp}}{\text{ihp}}$$

In a compressor, the *numerator* is the ideal work (or hp).

$$\text{(w)} \qquad \text{compressor efficiency } \eta_c = \frac{\substack{\text{work of steady flow} \\ \text{isentropic compression}}}{\text{shaft work input}} = \frac{W}{W_b}$$

Thermal Efficiencies

Thermal efficiency of a cycle is as given in equation (u).

For thermal efficiencies of engines, the energy chargeable against them, E_c, is:

(x) for gas turbine and internal combustion engine, $E_c = m_f q_l$

The units must be such that $e' = W'/E_c$ is dimensionless. m_f is the mass of fuel and q_l is the *lower heating value.* (Sometimes q_h, the *higher heating value* is used.)

For steam turbines and engines, $E_c = h_1 - h_{f2}$ BTU/lbm, where h_1 is the enthalpy of steam entering, and h_{f2} is the enthalpy of the liquid at the exhaust state, *except* where the engine is a part of a reheat-regenerative cycle. In this latter case, E_c = enthalpy of the steam entering the engine *plus* heat added during reheat processes *minus* the enthalpy of the liquid as it leaves the last feedwater heater. Actual thermal efficiencies depend on where the work or power is measured: indicated, $e_i = W_i/E_c$; brake, $e_b = W_b/E_c$; combined, $e_k = W_k/E_c$.

$$\text{boiler efficiency} = \frac{m_{bo}(h_{\text{out}} - h_{\text{in}})}{w_f q_h}$$

m_{bo} is lbm/hr of H_2O flowing through the system. h_{out} is the enthalpy of the departing steam. h_{in} is the enthalpy of the entering water. w_f is lbm/hr of fuel. q_h BTU/lbm = higher heating value of fuel. In general, boiler efficiency is the fraction of the higher heating value of the fuel that is transferred to the H_2O.

EXAMPLE: Tests of a 6-cylinder, 4-inch bore by $3\frac{1}{8}$-inch stroke aircraft engine at full throttle show bhp = 79.5 at 3400 rpm. The compression ratio is 8:1. The specific fuel consumption is 0.56 lbm/bhp-hr. The higher heating value of the fuel is 19,800 BTU/lbm. Determine (a) brake mean effective pressure, (b) brake torque, (c) brake thermal efficiency, (d) brake engine efficiency based on cold air standard if the ideal efficiency is $e = 1 - 1/r^{k-1}$, where r is the compression ratio.

SOLUTION: Note that the higher heating value is given even though the tendency would be to use the lower heating value.

(a) Use $pLAN$ equation.

$$\text{piston area } A = \frac{\pi D^2}{4} = \frac{\pi(16)}{4} = 12.57 \text{ in}^2$$

The number of power cycles per minute of a 6-cylinder, 4-stroke engine is

$$N = (6)\left(\frac{1}{2}\right)(3400) = 10{,}200 \text{ cpm}$$

$$\text{stroke } L = 3.125/12 = 0.2604 \text{ ft}$$

$$p_{mb} = \frac{33{,}000 \text{ bhp}}{LAN} = \frac{(33{,}000)(79.5)}{(0.2604)(12.57)(10{,}200)} = 78.6 \text{ psi} \quad (\textit{Ans.})$$

(b) Work is $M\theta$. Power is $M\omega$, where M is the moment acting through angular displacement θ radians and ω radians/second.

$$\omega = \frac{2\pi n}{60} = \frac{2\pi(3400)}{60} = 356 \text{ radians/second}$$

$$\text{bhp} = \frac{T_b \omega}{550 \text{ ft-lbf/sec-hp}} = 79.5 = \frac{356 T_b}{550}$$

From this, the torque is $T_b = 123$ ft-lbf. (*Ans.*) (Check units.)

(c)

$$E_c = m_f q_h = \left(0.56 \frac{\text{lbm}}{\text{bhp-hr}}\right)\left(19{,}800 \frac{\text{BTU}}{\text{lbm}}\right)$$

$$= 11{,}088 \text{ BTU/bhp-hr}$$

$$= (79.5)(11{,}088) \text{ BTU/hr}$$

$$W_b = \left(2544 \frac{\text{BTU}}{\text{hp-hr}}\right)(79.5 \text{ hp}) \text{ BTU/hr}$$

$$e_b = \frac{(2544)(79.5)}{(11{,}088)(79.5)} = 23\% \quad (\textit{Ans.})$$

(d) Cold air is air at approximately atmospheric temperature (i.e., $k = 1.4$).

$$e = 1 - \frac{1}{8^{1.4-1}} = 0.565$$

$$\eta_b = \frac{W_b}{W} = \frac{W_b/Q_A}{W/Q_A} = \frac{e_b}{e} = \frac{0.23}{0.565} = 40.7\% \quad (Ans.)$$

TWO-PHASE SYSTEM

In standard symbols, the properties of mixtures of liquids and vapors (one substance) are:

[LOW QUALITY]

$$v = v_f + xv_{fg}$$
$$h = h_f + xh_{fg}$$
$$s = s_f + xs_{fg}$$

[HIGH QUALITY]

$$v = v_g - yv_{fg}$$
$$h = h_g - yh_{fg}$$
$$s = s_g - ys_{fg}$$

Internal energy is $u = h - pv/J$. $x =$ the *quality*, or fraction by weight of mixture that is vapor, and y is the fraction of mixture that is liquid. y is more accurate when the quality is high. Handle processes as previously outlined. Avoid using equations that are valid only for ideal gases. Properties are obtained from tables.

MIXTURES OF GASES AND VAPORS

Dalton's law: Pressure p_m of a mixture in thermal equilibrium is

(y) $$p_m = p_A + p_B + p_C + \dots$$

This is the sum of the partial pressures of the components, A, B, C, and so forth. The *volumetric analysis* expresses the amounts of the components in a mixture by the percentages of the total volume that each component would occupy if the various gases were placed in separate compartments at the pressure p_m and temperature T_m of the mixture.

For ideal gases, the *volumetric fraction* of a component A is p_A/p_m, and so forth. Steam and some other imperfect gases qualify if their pressure is low (e.g., less than 1 psia).

The *gravimetric fraction* of a component A is ρ_A/ρ_m, called "percentage by weight" when in percentage form.

The density of a mixture is the sum of the densities of the constituents: $\rho_m = \rho_A + \rho_B + \rho_C + \ldots$.

Atmospheric Air

In figure 9.3,

$T_d = T_1 =$ dry-bulb temperature

$T_w =$ wet-bulb temperature (approximately the same as the temperature after an adiabatic saturation process)

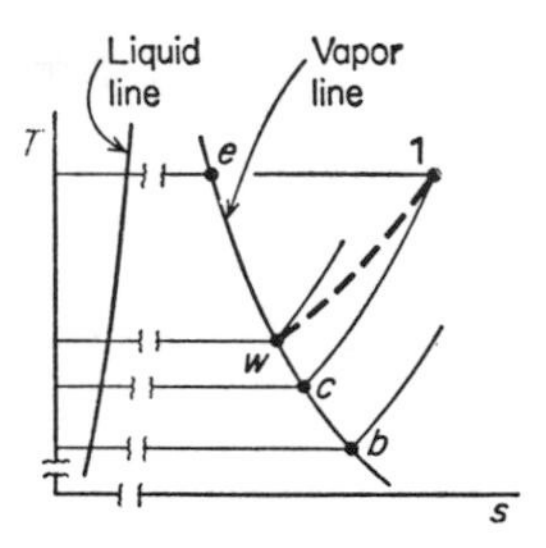

Figure 9.3 Curves such as $1c$ are constant pressure lines. $1w$ is the irreversible adiabatic saturation curve. c is the dew point $(p = C)$. The pressure at e is that corresponding to temperature $T_e = T_1$. $bcwe$ is the locus of all saturated vapor states.

The dew-point temperature, T_c, of atmospheric air is that at which the H_2O in the air begins to condense, after a constant total pressure cooling. It is the saturation temperature corresponding to the pressure p_{v1} of the steam.

The relative humidity, ϕ, of any superheated vapor is the actual pressure of the vapor *divided by* the saturation pressure corresponding to the actual temperature of the vapor. In figure 9.3, this is p_1/p_e for state 1.

(z)
$$\phi = \frac{p_{v1}}{p_{ve}} \approx \frac{\rho_{v1}}{\rho_{ve}} = \frac{v_{ve}}{v_{v1}}$$

Subscript v is added to emphasize *vapor*, the component that can condense. The components that do not condense for ordinary temperatures are collectively called the *dry gases*, or *dry air* for atmospheric air. p_a = pressure of dry air, ρ_a = density of dry air, and so forth.

$$\text{humidity ratio } \omega = \frac{\rho_v}{\rho_a} \qquad \text{in units of } \frac{\text{lbm vapor}}{\text{lbm dry air}}$$

$$\rho_v = \frac{p_v}{R_v T_v}$$

$$\rho_a = \frac{p_a}{R_a T_a}$$

When the vapor pressure is low,

$$p_m = p_v + p_a$$

Humidity ratio is also called *specific humidity*, often expressed in grains (7000 grains/lbm).

If atmospheric air (or other gas with a condensing or evaporating component) undergoes a process during which H_2O (or other component) evaporates or condenses, simply set up an equation for the energy flow considering each component separately, but have the denominator of each term the same, e.g., "per lbm of dry air." To illustrate, the energy diagram or control volume of a cooling tower (for cooling water) or air cooler (water cools air) is shown in figure 9.4. The energy balance is

$$h_{a1} + \omega_1 h_{v1} + m_A h_{fA} = h_{a2} + \omega_2 h_{v2} + m_B h_{fB}$$

h_a is the enthalpy of dry air (BTU/lbm of dry air). h_v BTU/lbm H_2O is the enthalpy of steam in air. h_f BTU/lbm H_2O is the enthalpy of water. w lbm H_2O/lbm dry air is mass of water. The mass balance (H_2O in = H_2O out) for the H_2O is

$$\omega_1 + m_A = \omega_2 + m_B$$

State 2 is often assumed to be "saturated air" in a cooling tower.

EXAMPLE: Calculate the vacuum in a condenser when there is 0.11 lbm of dry air per pound of steam. The barometer reads 30 inches Hg. The temperature in the condenser is 92 °F.

SOLUTION: Steam tables indicate that partial pressure of vapor (at 92 °F) is $p_v = 0.7432$ psia.

$$\rho_v = 1/v_v = 1/441.3 \text{ lbm/cu ft}$$

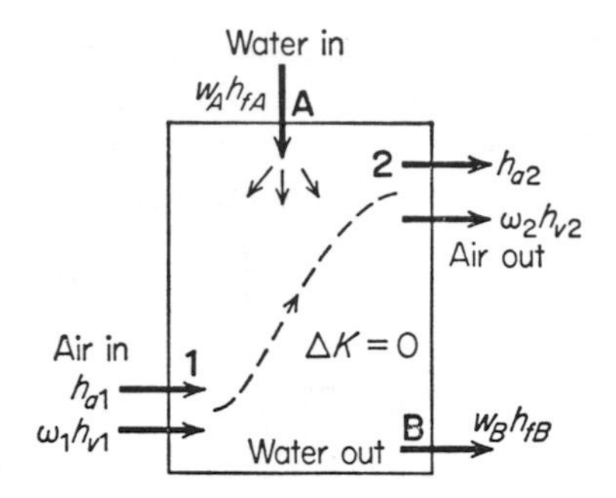

Figure 9.4 Cooling tower.

Let p_a = partial pressure of dry air in condenser. Then, the density of dry air is $\rho_a = p_a/R_a T_a$. The more accurate solution is by ratio of the foregoing densities. But, at low pressure, the steam acts similarly to an ideal gas. Use $p = \rho RT$:

$$0.11 \frac{\text{lbm dry air}}{\text{lbm v}} = \frac{\rho_a}{\rho_v} = \frac{p_a R_v T_v}{R_a T_a p_v} = \frac{p_a R_v}{p_v R_a}$$

In the above equation,

$$\begin{aligned} T_v &= T_a \\ R_v &= 1545/M = 1545/18.016 = 85.7 \\ R_a &= 53.3 \\ p_v &= 0.7432 \end{aligned}$$

Solving for p_a,

$$\begin{aligned} p_a &= 0.0508 \text{ psia} \\ p_m &= p_v + p_a = 0.7432 + 0.0508 = 0.794 \text{ psia} \\ p_m &= \frac{0.794}{0.491} = 1.62 \text{ inches Hg} \end{aligned}$$

$$\text{vacuum (gauge pressure)} = 30 - 1.62 = 28.38 \text{ inches Hg} \quad (Ans.)$$

REVERSED CYCLES

Reversed cycles are used for both refrigeration and heating (popularly called *heat pump*). Comparative performance is given by the *coefficient of performance*, γ.

$$\gamma_r = \frac{\text{refrigeration}}{\text{work}} = \frac{Q_A}{W} \qquad \text{[USED FOR COOLING]}$$

$$\gamma_h = \frac{\text{heating effect}}{\text{work}} = \frac{Q_R}{W} \qquad \text{[FOR WARMING]}$$

Capacity of refrigerating cycles is stated in *tons* of refrigeration (1 ton = 200 BTU/minute of cooling). If N = tons of refrigeration, total refrigeration is $Q_A = 200N$ BTU/minute. If the corresponding work is represented by hp, this is equivalent to $W = 42.4\text{hp}$ BTU/minute.

$$\gamma_r = Q_A/W = 200N/(42.4\text{hp}) = 4.72N/\text{hp}$$

Most operating cycles are vapor (compressions) cycles. Heat added occurs in an *evaporator*. Heat rejected occurs in a *condenser*. Liquid refrigerant throttles through an expansion valve from high to low pressure ($\Delta h = 0$).

EXAMPLE: A Freon-12 vapor refrigeration cycle operates with a condenser temperature of 120 °F and an evaporator temperature of −20 °F. The liquid refrigerant enters the expansion valve with an enthalpy of 31.16 BTU/lbm. The enthalpy of the vapor at the entrance and exit of the compressor is 77.32 BTU/lbm and 104.52 BTU/lbm, respectively. The rate of refrigeration is 1 ton. Determine (a) the horsepower to compress the vapor, (b) the rate of heat transfer to the condenser cooling water in BTU/hour.

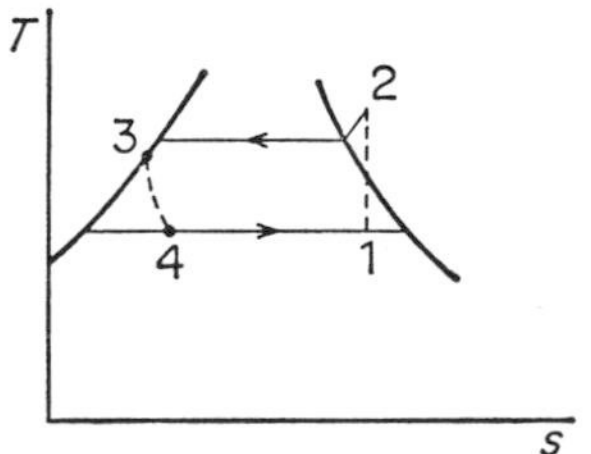

Figure 9.5 Refrigeration cycle.

SOLUTION: (a) See figure 9.5. The location of line 1–2 is unknown, but $h_1 = 77.32$, $h_2 = 104.52$, and $h_3 = h_4 = 31.16$ BTU/lbm. The flow through the evaporator is steady. The refrigeration is

$$Q_A = h_1 - h_4 = 77.32 - 31.16 = 46.16 \text{ BTU/lbm}$$

Mass of refrigerant needed is

$$m = \frac{200 \text{ BTU/minute}}{46.16 \text{ BTU/lbm}} = 4.33 \text{ lbm/minute}$$

Flow through the compressor is steady, so

$$W = h_2 - h_1 \text{ BTU/lbm} \quad \text{(a positive number)}$$

$$\text{hp} = \frac{w(h_2 - h_1)}{42.4} = \frac{(4.33)(104.52 - 77.32)}{42.4} = 2.78 \text{ hp} \quad (\textit{Ans.})$$

(b) Heat in condenser (steady flow) is

$$Q_R = h_2 - h_3 \text{ BTU/lbm}$$
$$Q_R = w(h_2 - h_3)(60) = (4.33)(104.52 - 31.16)(60)$$
$$= 19{,}050 \text{ BTU/hr.} \quad (\textit{Ans.})$$

PRACTICE PROBLEMS

9.1. A completely air-tight 300-cm cylinder is divided into two parts by a freely moving, air-tight, heat-insulating piston. When the temperatures in the two compartments are each equal to 27 °C, the piston is located 100 cm from one end of the cylinder. How far will the piston move if the gas in the smaller part of the cylinder is heated to 74 °C, and the temperature of the larger section remains constant? Assume perfect gas laws hold, and neglect the thickness of the piston.

Ans. 9.93 cm

9.2. A mixture of 60% nitrogen, 10% carbon dioxide, and 30% hydrogen by volume is heated from 40 °F to 250 °F at a constant pressure of 1.0 atmosphere. The volume of the total mixture is 5.0 cu ft at 40 °F. (a) What is the total weight of the mixture? (b) What is the specific heat of the mixture? (c) How much energy is added to the mixture? (d) How much energy is available for external work during the process?

PHYSICAL PROPERTIES	SPECIFIC HEAT	GAS CONSTANT
	$c_p \frac{\text{BTU}}{\text{lbm-°F}}$	$R \frac{\text{ft-lbf}}{\text{lbm-°R}}$
nitrogen	0.2485	55.3
carbon dioxide	0.2175	35.0
hydrogen	3.140	762.0

Ans. (a) 0.299 lbm, (b) 0.322 BTU/lbm-°R. (c) 20.3 BTU, (d) 4450 ft-lbf

9.3. The specific heat of an ideal gas at constant pressure is 0.2025 BTU/lbm-°F and at constant volume, 0.1575 BTU/lbm-°F. What is the final volume of 10.0 cu ft of gas at an initial pressure of 25 inches Hg and a final pressure of 5 atm? (Mercury weighs 0.49 lbm/in^3.)

Ans. 1.67 cu ft (for $T = C$), 2.49 cu ft (for $s = C$)

9.4. In a certain chemical process, two liquids enter a mixing chamber and are thoroughly mixed before being discharged at 80 °F and 50 gpm. Liquid (1) enters at 140 °F and has a specific heat of 10 BTU/gal-°F. Liquid (2) enters at 65 °F and has a specific heat of 8.33 BTU/gal-°F. Assume that there is no chemical reaction and that there is no heat lost or gained by the system. (a) What is the flow for liquid (1) and for liquid (2)? (b) What is the specific heat of the mixed liquid?

Ans. (a) 8.6 gpm, 41.4 gpm, (b) 8.62 BTU/gal-°F

9.5. A centrifugal compressor pumps 100 lbm/minute of air from 14.7 psia and 60 °F to 50 psia and 270 °F by an irreversible process. The temperature rise of 50 lbm/minute of circulating water about the casing is 12 °F. What horsepower is required? Neglect any changes in kinetic energy. *Note*: 2545 BTU/hr= 1 hp. Assume no heat is given up to the surrounding atmosphere.

Ans. 133 hp

9.6. An oxygen cylinder of volume 2.3 cu ft has a pressure of 2200 psig and is at 70 °F. (a) What weight of oxygen is in the cylinder? (b) What is the gauge pressure in the cylinder if the temperature is raised to 120 °F?

Ans. (a) 28.7 lbm, (b) 2405 psig

9.7. Heat is supplied to 20 lbm of ice at 32 °F at the rate of 160 BTU/sec. If the heat of fusion is 144 BTU/lbm, how long will it take to convert the ice to water at 50 °F?

Ans. 20.25 sec

9.8. The volume in the cylinder of a one-cylinder air compressor is 0.57 cu ft at the beginning of the compression stroke with air at atmospheric pressure. The piston compresses the air polytropically to 69.8 psig according to the law $PV^{1.35}$ = constant. What is the volume under compression at this pressure?

Ans. 0.157 cu ft

9.9. If an ideal gas is compressed from a lower pressure to a higher pressure at constant temperature, which of the following is true? (a) The work required will be zero. (b) The volume remains constant. (c) The volume will vary inversely as the absolute pressure. (d) The volume will vary directly as the gauge pressure. (e) Heat is being absorbed by the ideal gas during the compression.

Ans. (c)

9.10. A steam turbine uses 640 lbm/hr of 235.3 psig steam with 20 °F super-heat. The turbine exhaust steam pressure is 25.3 psig and in a saturated state. (a) How much power (in BTU/hr) is given up by the steam as it passes through the turbine? (b) What is the magnitude of the volume (in cu ft/sec) begin exhausted at the exhaust conditions listed above? (c) How much heat (in BTU/lbm) is given up by the exhaust steam if it is reduced from 25.3 psig to a saturated liquid at atmospheric pressure?

Ans. (a) 28,877 BTU/hr, (b) 1.867 cu ft/sec, (c) 989.6 BTU/lbm

9.11. A small steam-generating plant burns oil as a fuel. The oil has a specific gravity of 1.008 and costs $2.50 per barrel. The heating value of the oil is 18,250 BTU/lbm. The boiler operates at 75% thermal efficiency when taking in feedwater at 180 °F and delivering dry saturated steam at 100 psig. What is the fuel cost per 1000 lbm of steam on this basis? 1 barrel = 42 gallons.

Ans. $0.54

9.12. Relative humidity can be defined as 100 times the ratio of: (a) the weight of moisture per cubic foot of dry air, (b) the actual vapor pressure to the pressure of saturated vapor at the same temperature, (c) the wet-bulb temperature divided by the dry-bulb temperature.

Ans. (b)

9.13. In the continuous process of imparting a black oxide surface coating on steel couplings, the couplings are submerged in a molten salt bath for 8 minutes and then removed to a water rinse bath. The rinse bath removes the salt that has adhered to the couplings and cools them to the proper temperature for handling. A cold water supply and overflow to the rinse tank provides the necessary control for maintaining the proper rinse temperature. The rinse bath is thoroughly agitated at all times. Operating data are as follows: weight of steel coupling, 5.5 lbm; specific heat of steel, 0.11 BTU/lbm-°F; temperature of salt bath, 450 °F; temperature of rinse bath, 125 °F; temperature of cold water supply, 71 °F; amount of rinse water maintained in bath, 1.25 cu ft; amount of salt carried over to rinse bath by one coupling, 0.013 lbm/coupling; production rate, 30 couplings/hr. (a) At what rate is heat (BTU/hr) being added to the rinse water by the quenching of the couplings? (b) At what rate (lbm/hr) must cold water be supplied to the rinse bath to maintain 125 °F? (c) What is the salt concentration (lbm of salt per lbm of rinse solution) in the rinse tank after operations have been in production long enough to make the rinse bath salt concentration essentially constant?

Ans. (a) 5900 BTU/hr, (b) 109 lbm/hr, (c) 0.00357 lbm salt/lbm water

9.14. Fifty-four gallons of diesel fuel are burned in an engine. How many hp-hours will be generated at the piston if the thermal efficiency is

30%? The density of the fuel oil is 7.8 lbm/gal. Its lower heating value is 18,000 BTU/lbm. 1 BTU = 778 ft-lbm.

Ans. 894 hp-hr

9.15. If 1000 lbm of air at 95 °F dry bulb and 78 °F wet bulb is cooled to 55 °F and 85% relative humidity, find: (a) total heat removed, (b) final wet-bulb temperature, (c) sensible heat removed, (d) latent heat removed, (e) initial dew-point temperature, (f) final dew-point temperature, (g) the weight of moisture condensed from air.

Ans. (a) 19,080 BTU, (b) 52.5 °F, (c) 9600 BTU, (d) 9480 BTU, (e) 71.7 °F, (f) 50.5 °F, (g) 8.95 lbm

9.16. An automobile tire is inflated to 32 psig pressure at 50 °F. After the car has been operated, the temperature rises to 75 °F. Assuming that the volume remains constant, what is the final gauge pressure?

Ans. 34.4 psig

9.17. A rock weighs 1200 lbm in air, 700 lbm in water. What is the volume of the rock in cubic inches?

Ans. 13,850

9.18. One pound of a mixture is composed of 90% hydrogen and 10% water vapor by volume. The total pressure of the mixture is 10 psia and its temperature is 300 °F. The mixture is cooled to 40 °F, while the total pressure remains constant. Find: (a) the dew point for the mixture, (b) the weight of H_2O condensed.

Ans. (a) 101.7 °F, (b) 0.442 lbm

9.19. Three pounds of nitrogen expand in a nonflow polytropic process. During the process, the entropy increases by 0.2 BTU/lbm-°R. The temperature increases by 200 °F during the process. The initial temperature is 100 °F. Find (a) heat transfer (including direction), (b) internal energy change (including direction).

Ans. (a) +393 BTU, (b) +106 BTU

9.20. Two pounds of hydrogen are cooled at constant volume from 300 °F to 200 °F. Heat from the hydrogen is transferred to 30 lbm of oxygen, originally at 60 °F. The oxygen is heated at constant pressure. With respect to a refrigerator or sink temperature of 30 °F, find the change in available energy that occurs because of the over-all heat transfer.

Ans. 94 BTU

9.21. Nitrogen acting as a perfect gas expands adiabatically in a nozzle from a pressure of 200 psia and a temperature of 300 °F. The pressure at the exit section is 25 psia. The nozzle efficiency is 80%. Flow rate is 2 lbm/sec. Find the area at the exit section in square inches if the velocity of approach is zero.

Ans. 1

9.22. The volume of an automobile tire is 1.2 cu ft. How many such tires can be inflated to a pressure of 30 psig from an air storage tank with a volume of 100 cu ft, if the air in the tank is originally at 200 psig and 70 °F? The barometer reads 15 psia. Assume that the temperature of the air does not change and the tube in the tire is collapsed when inflation begins.

Ans. 315

9.23. A single-cylinder reciprocating air compressor has a clearance of 7%. The suction pressure is 20 psia and the discharge pressure is 100 psia. Suction temperature is 100 °F. Atmospheric temperature is 70 °F. Compression is considered to be isentropic. The displacement volume of the compressor cylinder is 2.0 cu ft. What volume of air measured at suction conditions enters the cylinder per stroke of the compressor?

Ans. 1.7 cu ft

9.24. An ammonia compressor is used in a heat pump cycle. The suction pressure is 30 psia. The discharge pressure is 160 psia. Saturated liquid ammonia enters the throttle valve. The refrigerating effect is 500 BTU/lbm of ammonia. Find the coefficient of performance as a heating cycle.

Ans. 5.36

9.25. A steam power plant operates on the regenerative cycle. Steam enters the turbine at 600 psia and 700 °F. It expands isentropically to 200 psia where steam is bled to a feedwater heater. The exhaust pressure of the turbine is 2.0 psia. The weight of steam bled is 0.10 lbm per lbm of throttle steam. Find: (a) the enthalpy of exhaust steam (BTU/lbm), (b) the temperature of discharge from the feedwater heater, (c) the thermal efficiency of the cycle.

Ans. (a) 922 BTU/lbm, (b) 240 °F, (c) 34.8%

9.26. (a) How many BTU are required to raise 10 lbm of ice at 0 °F to steam at 212 °F? (b) What is the weight of a cubic foot of air when the pressure is 50 psia and the temperature is 180 °F? (c) A quantity of air at

60 °F under a pressure of 14.7 psia has a volume of 5 cu ft. What is the volume of the same air when the temperature is changed to 120 °F at a constant pressure?

Ans. (a) 13,097 BTU, (b) 0.211 lbm, (c) 5.58 cu ft

9.27. The atmospheric air has a dry-bulb temperature of 87 °F and a wet-bulb temperature of 65 °F. Calculate: (a) the specific humidity in grains per pound of dry air, (b) the dew-point temperature, (c) the relative humidity. Barometric pressure is 14.7 psia.

Ans. (a) 56.7, (b) 52 °F, (c) 30%

9.28. A steam turbine carries a load of 32,000 kw and uses 360,000 lbm/hr of steam. The engine efficiency is 80%, its exhaust steam is at 1 inch Hg abs., and it has an enthalpy of 950 BTU/lbm. What are the temperature and pressure of the steam as it enters the turbine?

Ans. 465 °F, 170 psia

9.29. The efficiency of the engine and generator in a 500-kw generating set is 85%, the steam pressure is 150 psia, and the feedwater temperature is 180 °F. The engine uses 20 lbm/ihp-hr of steam. Evaporation from and at 212 °F is 10 lbm of water per lbm of dry coal. Coal contains 13,000 BTU/lbm. What is the thermal efficiency of the plant?

Ans. 7.7%

9.30. An engine operating on the Carnot cycle uses 10 lbm/sec of steam as the working substance. During the heat addition process, the steam goes from a saturated liquid to a saturated vapor state. Heat addition takes place at 200 psia and heat rejection at 15 psia. (a) Sketch the cycle to pv and Ts coordinates and show the saturation lines. Determine (b) the heat addition (BTU/sec) (c) the heat rejection (BTU/sec), (d) the net hp output, (e) the pumping hp input, (f) the thermal efficiency.

Ans. (b) 8430 BTU/sec, (c) 6730 BTU/sec, (d) 2400 hp, (e) 280 hp, (f) 20.1%

9.31. One pound of air completes a reversible cycle consisting of the following three processes: (1) From a volume of 2 cu ft and temperature of 40 °F, the air is compressed adiabatically to half the original volume. (2) Heat is added at constant pressure until the original volume is reached. (3) The air is returned reversibly to its original state. (a) Show the cycle on pv and Ts coordinates. (b) Calculate the heat transfer, work, and internal energy change for each process, and show *signs* of results.

Ans. (b) (1) 0 BTU, −21,400 ft-lbf, 27.4 BTU, (2) 158.4 BTU, 35,200 ft-lbf, 113 BTU, (3) −141 BTU, 0, −141 BTU

9.32. One pound of saturated steam at 400 °F expands isothermally to 60 psia. Determine: (a) the change of entropy, (b) the heat transferred, (c) the change of enthalpy, (d) the change of internal energy, (e) the work.

Ans. (a) 0.1863 BTU/lbm-°R, (b) 160.3 BTU, (c) 32.6 BTU/lbm, (d) 25 BTU, (e) 105,000 ft-lbf

9.33. One pound of steam at 400 psia and 600 °F expands isentropically to a final temperature of 200 °F. Determine: (a) the initial enthalpy, entropy, volume, internal energy; (b) the final pressure, quality, volume, enthalpy, internal energy; (c) the heat transferred; (d) the change of internal energy; (e) the work done during the process for both a steady flow and a nonflow process. (f) Sketch on the temperature-entropy plane, showing saturated liquid and vapor lines.

Ans. (a) 1306.9 BTU/lbm, 1.5894 BTU/lbm-°R, 1.477 ft^3, 1197.6 BTU, (b) 11.526 psia, 87.4%, 29.4 ft^3, 1022.7 BTU/lbm, 960 BTU, (c) 0, (d) 237.6 BTU, (e) 237.6 BTU (nonflow), 284.2 (steady flow) BTU

9.34. Ten pounds of air are at a barometric pressure of 30.55 inches Hg and at a temperature of 60 °F as the *initial* conditions for *each* of the three separate processes: (a) If the air is heated at constant volume to 1485 psig, what is the final temperature in °F? (b) If the air is compressed at constant temperature to 1485 psig, what is the final volume in cubic feet? (c) If the air is cooled at constant pressure until the temperature is −330 °F, what will be the final volume in cubic feet?

Ans. (a) 51,540 °F (assuming it still exists), (b) 1.28 ft^3, (c) 32.1 ft^3

9.35. A 2500-kw steam turbine is operated at its rated capacity under the following conditions: barometric pressure, 14.7 psi; steam pressure at throttle, 201.3 psig; steam temperature at throttle, 490.0 °F; steam exhaust pressure, 2 inches Hg abs.; throttle flow (lbm steam/hr), 33,500. Compute the following: (a) the theoretical steam rate of this turbine, (b) its engine efficiency, (c) its actual thermal efficiency, (d) its heat rate.

Ans. (a) 7.03 lbm/hp-hr, (b) 70.3%, (c) 21.3%, (d) 11,920 BTU/hp-hr

9.36. A turbine operates with its steam supply at a pressure of 214 psia and a temperature of 400 °F and exhaust pressure at 2 inches Hg abs. The output as measured by means of a dynamometer is 500 hp and the steam consumption is 8200 lbm/hr. Calculate: (a) the thermal efficiency, (b) the engine efficiency.

Ans. (a) 13.6%, (b) 45.2%

9.37. One-half of a molecular weight of a perfect gas exists at a pressure of 100 psia. A heat transfer of 200 BTU at a constant volume causes the temperature to change by 100 °F. Find the specific heat per mole at constant pressure.

Ans. 5.98 BTU/mol-°R

9.38. A steam turbine receives 5000 lbm/hr of steam at 100 fps velocity and an enthalpy value of 1500 BTU/lbm. The exhaust steam leaves at 800 fps velocity and enthalpy value of 900 BTU/lbm. Find the horsepower input to the turbine blades.

Ans. 1154 hp

10 *FLUID MECHANICS*

SYMBOLS

A = area
a = acceleration; sonic velocity
C = a constant
D = diameter
d = depth of liquid in channel
E = bulk modulus; energy
F = force
f = friction factor, pipe flow
G = mass rate of flow per unit area
g = acceleration due to gravity (ft/sec^2)
g_c = standard gravitational constant (32.2. lbm-ft/lbf-sec^2)
H = total head, ft
h = enthalpy; depth of fluid
h_f = friction head loss
K = kinetic energy
L = length
M = molecular weight
m = mass in slugs ($\text{lbm}/g_c = W/g$)
n = roughness coefficient
P = potential energy of gravitation
p = unit pressure, lbf/ft^2
Q = heat
R = hydraulic radius; gas constant
$\mathbf{R}$ = Reynolds number
S = slope
s = entropy
T = absolute temperature, °R
u = specific internal energy
V = total volume; volume of flow, cu ft/sec
v = specific volume
$\mathcal{V}$ = speed, ft/sec

W = weight, lbf; shaft work
z = elevation (ft) of body with potential energy P
γ = specifc weight, lbf/cu ft
μ = absolute viscosity
ν = kinematic viscosity
ρ = density (lbm/ft^3)
τ = shear stress

HYDROSTATICS

Specific gravity of a liquid is the ratio of the weight (or mass) of the liquid divided by the weight (or mass) of a like (unit) volume of water at atmospheric pressure and some standard temperature, say, 60 °F or 68 °F. For gases, the reference mass is air.

Specific weight, γ, is the weight of a unit volume of material, usually lbf/cu in. or lbf/cu ft. For water, $\gamma \approx 62.4$ lbf/cu ft on the earth's surface.

Pressures are measured by columns of liquid, as well as in other ways. If a column of liquid whose *specific weight* is γ lbf/cu in. stands vertically h inches high with a vacuum on top, as in an inverted tube barometer, the pressure exerted by the liquid column is its weight divided by the constant area A in^2 over which it acts:

(a) $$p = \frac{\gamma V}{A} = \frac{\gamma A h}{A} = \gamma h \text{ psi}$$ [LIQUID COLUMN]

(γ lbf/cu in. $\times$ h inches = psi.) For example, the pressure at the bottom of a tank of water, 13 feet deep, is

$$p = (62.4)(13) = 811 \text{ lbf/ft}^2 = 5.62 \text{ psi}$$

Atmospheric or other pressure, if any, on top of the water, is then added. If the tube is inclined, h is always the vertical distance from the top of liquid to the level where the pressure is desired.

Liquid pressure, called *hydrostatic pressure*, varies linearly with h. Total pressure is $p = p_o + \gamma h$ if p_o is the pressure on the surface of liquid. In figure 10.1(a), the resultant hydrostatic force F on the *rectangular* side of the container passes through the centroid of the triangle formed by this pressure distribution. The point on the area at depth h_c through which F passes is called the *center of pressure*. The "average" pressure occurs at the centroid of the area A (any shape) over which the pressure acts, depth $\bar{h}$. Hence, total force, if surface pressure is relatively small, is

(b) $$F = pA = \gamma \bar{h} A \text{ lbf}$$

Fluid pressure always acts normal to the area of the surface it contacts. For an area of any shape submerged in any position, whose trace is BC in figure 10.1(b) and (c), the resultant normal force is $F = \gamma\overline{h}A$, where $\overline{h}$ is the vertical distance to the centroid of the area A. To locate the line of action of this force, take moments about an axis at the liquid surface, manipulate, and find

$$h_c = \overline{h} + \frac{\overline{I}}{\overline{h}A} \tag{c}$$

$\overline{I}$ is the moment of inertia of the area A about a centroidal axis parallel to the fluid surface. h_c locates the center of pressure, as in figure 10.1.

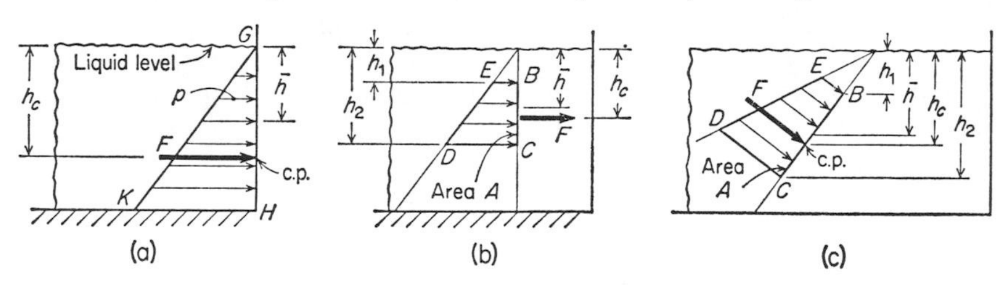

Figure 10.1 Rectangular surface

If the surface A is a rectangle, the line of action of the resultant fluid pressure passes through the centroid of the area $BCDE$, as in figure 10.1(b) and (c). If the density of the fluid varies, the basic equation is

$$\int dp = \int \gamma\, dh$$

h is positive in the direction of measuring depth. In figure 10.1(c), to get the horizontal force on area BC, use the horizontally projected area. Where the submerged "area" has the same fluid pressing on both sides, as in figure 10.1(b) and (c), the fluid forces on opposite sides are numerically the same.

EXAMPLE 1: A triangular channel, vertex down, has a top width of 8 feet and a depth of 6 feet. A bulkhead in the channel has water to the top on one side only. Find the force on the bulkhead and the location of the force.

SOLUTION: Make a sketch to help to follow the solution. The centroid of the triangle is one-third of the height from a base, $\overline{h} = 6/3 = 2$ ft, $A = (8)(6)/2 = 24$ ft^2, $\gamma = 62.4$ lbf/cu ft.

$$F = \gamma\overline{h}A = (62.4)(2)(24) = 2995 \text{ lbf} \quad (Ans.)$$

This is the atmospheric pressure on the water surface *and* the bulkhead. It is presumed that the net force, as shown, is desired.

$$\overline{I} = \frac{bh^3}{36} = \frac{(8)(216)}{36} = 48 \text{ ft}^4$$

$$h_c = \overline{h} + \frac{\overline{I}}{\overline{h}A} = 2 + \frac{48}{(2)(24)} = 3 \text{ ft} \quad (Ans.)$$

EXAMPLE 2: In a hydraulic jack, a piston of 6-inch diameter is activated by a 2-inch piston. For a 9-lbf force on the smaller piston, what is the force on the larger piston?

SOLUTION: Fluid pressure is assumed the same at all points of the fluid. In the small piston, $p = F_s/A = 9/\pi$. On the large piston,

$$F_1 = pA = \frac{9}{\pi} \times \pi \times 9 = 81 \text{ lbf} \quad (Ans.)$$

INCOMPRESSIBLE FLOW

Euler and Bernoulli Equations

For any fluid in steady flow, the laws of conservation of energy and of mass apply. Writing for 1 pound in terms of differentials, we have

(d) $$dQ = du + d(pv) + dK + dP + dW$$

Use units as convenient, say ft-lbf/lbm. Then for 1 pound (lbm), $P = z$ ft-lbf/lbm, $K = \mathcal{V}^2/(2g_c)$, $v = 1/\rho$, $g =$ constant. Apply (d) in the general case of flow through a pump or hydraulic turbine, and so forth.

For example, let a hydraulic turbine operate under ideal (no loss) conditions and a head of z feet, that is $z_2 = 0$, $z_1 = z$, $Q = 0$, $\Delta u = 0$, $v = 1/\rho =$ constant (incompressible), $K_1 = 0$, and $\Delta p = 0$ if the pressure is atmospheric in both states 1 and 2. Hence,

$$W = z_1 - K_2 \text{ ft-lbf/lbm}$$

State 1 is some point on the surface of the reservoir and state 2 is at the point of discharge of the turbine (or tail-race level). The same numerical answer would be obtained if the water in the hydraulic turbine were taken as the system and the energy equation applied as in thermodynamics.

When flow through a "work machine" is not involved, $W = 0$. Rearranging (d),

$$\text{(e)} \qquad \frac{dp}{\rho} + \frac{d\mathcal{V}^2}{2g_c} + dz = dQ - du - p\,dv$$

The right-hand side includes the thermal effects of fluid friction. Both Euler and Bernoulli equations are derived from mechanics considerations and Newton's laws, but the end results are the same as (e) when:

For Euler's,

$$dQ - du - p\,dv = \frac{\tau\,dx}{\rho R}$$

R = hydraulic radius, and the right-hand side is the friction term (shear of fluid).

For Bernoulli's,

$$dQ - du - p\,dv = 0$$

This is for frictionless flow. For incompressible flow (for liquids and for gases with small changes of v), assume $dv = 0$. If the flow is nearly adiabatic, $Q \approx 0$, in which case the frictional effect appears entirely as a change of internal energy. Many applications are made with a "head-loss" term h_f for the frictional effects, in which case the integrated form of (e) in its most applicable form is

$$\text{(f)} \qquad \frac{p_2}{\rho_2} - \frac{p_1}{\rho_1} + \frac{\mathcal{V}_2^2}{2g_c} - \frac{\mathcal{V}_1^2}{2g_c} + z_2 - z_1 = -h_f \text{ ft-lbf/lbm}$$

If a "pound force" is cancelled by a "pound mass" (ordinarily not advisable), each term in (f) can be and is interpreted as a *head* in feet: p/ρ = *pressure head*, $\mathcal{V}^2/(2g_c)$ = *velocity head*, z = *gravitational head*, and the friction head $h_f = 0$ for frictionless flow.

$$\text{(g)} \qquad \text{hydraulic radius } R = \frac{\text{cross-sectional area}}{\text{wetted perimeter}}$$

Do not count the free surface in the wetted perimeter.

Torricelli's Equation for Ideal Speed of Efflux

The kinetic energy of efflux, shown in figure 10.2, is equal to the liquid head, or $\mathcal{V} = \sqrt{2g_c h}$. In (f), this is the same as assuming a large reservoir

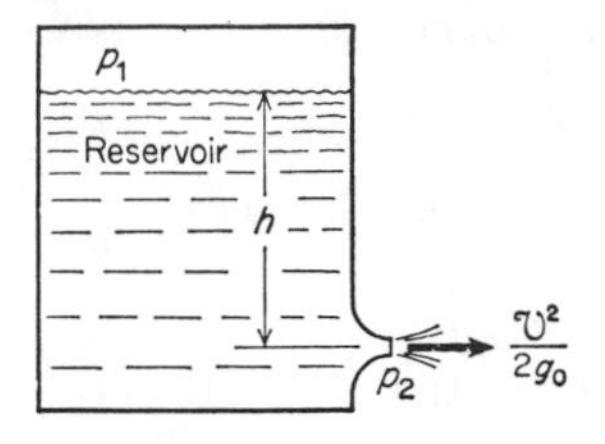

Figure 10.2

or $\mathcal{V}_1 = 0$. p/ρ is constant because $p_1 = p_2 =$ (say) atmospheric pressure, and $h_f = 0$.

EXAMPLE: A pipe carries water from A to point B, which is 6 feet higher in elevation than A. The diameter at A is 18 inches and at B, 27 inches. The pressure at A is 10 psi and at B, 10.5 psi. When the flow is 30 cu ft/sec, find the head loss and the direction of flow.

SOLUTION: Make a sketch showing data given. The problem says water goes from A to B and then asks for direction of flow—but there might be a slip here (or intentional misstatement). Find out for sure by computing total heads at A and B. *Flow is always in the direction of decreasing total head.* The kind of pressure certainly should be stated, but most water pipes are surrounded by the atmosphere, so pressures given are probably gauge, psig. Fluid speeds at A and B are, from $A\mathcal{V} = V = 30$ cu ft/sec and $A = \pi r^2$,

$$\mathcal{V}_A = \frac{30}{\pi 81/144} = 17 \text{ ft/sec}$$

$$\mathcal{V}_B = \frac{30}{\pi(13.5)^2/144} = 7.55 \text{ ft/sec}$$

At A, $z_a = 0$, and, the total head H_A is

$$H_A = \frac{p_A}{\rho_A} + \frac{\mathcal{V}_A^2}{2g_c} + z_A = \frac{(24.7)(144)}{62.4} + \frac{(17)^2}{64.4} = 61.5 \text{ ft}$$

$$H_B = \frac{(25.2)(144)}{62.4} + \frac{(7.55)^2}{64.4} + 6 = 65.1 \text{ ft}$$

Flow is from B to A with head loss $h_f = 65.1 - 61.5 = 3.6$ ft. (*Ans.*)

Incompressible Flow through Measuring Devices

A gas undergoing only a small expansion or compression has a small enough change of volume that the flow can be considered as incompressible. For gases (and for liquids) passing through venturis, flow nozzles, plate orifices, and against pitot tubes, the steady-flow energy equation, ith $Q = 0$ (system is well-insulated, $\Delta T \approx 0$, or time is small), $W = 0$ (no engine to drive or be driven), $\Delta u = 0$ (frictionless flow), $K_1 = 0$ (properties taken at stagnation state), $\Delta P = 0$ (for horizontal flow or insignificant change of elevation), and with $dv = 0$ (incompressible), becomes

(h)
$$p_2 v_2 - p_1 v_1 + K_2 = 0$$
$$\frac{\mathcal{V}_2^2}{2g_c} = v(p_1 - p_2) = \frac{p_1 - p_2}{\rho}$$

$v = v_1$ might be obtained from $p_1 v_1 = RT_1$ for an ideal gas. The correction factors for initial velocity ($\mathcal{V}_1$ is never absolutely zero) between sections 1 and 2 are:

$$I = \frac{1}{\sqrt{1 - (A_2/A_1)^2}} = \frac{1}{\sqrt{1 - (D_2/D_1)^4}} \qquad \text{[LIQUID]}$$

$$I = \frac{1}{\sqrt{1 - (A_2 v_1/A_1 v_2)^2}} \qquad \text{[GAS]}$$

That is, from (h),

$$\mathcal{V}_2 = I\sqrt{2gv(p_1 - p_2)} = I\sqrt{2g(p_1 - p_2)/\rho}$$

This is valid when $\mathcal{V}_1$ is significant and p_1 is not the stagnation pressure. If a liquid undergoes a material change of z, keep this term in the equation. The quantity of flow is found from thc continuity of mass equation

(i)
$$\frac{A_1 \mathcal{V}_1}{v_1} = \frac{A_2 \mathcal{V}_2}{v_2}$$

Units are, say, lbm/sec.

Apply a coefficient of discharge η_d to get the actual flow $V' = \eta_d V$. For venturi and flow nozzle, η_d is high (96% to 98%). For orifice, η_d is lower and more varied. See also *nozzles*. In thin plate orifices, the overall value of η_d includes the coefficient of contraction C_c (at the vena contracta). $\eta_d = C_v C_c$, where C_v is the velocity coefficient. The total volume of flow $V = \dot{m} g_c v$ cu ft/sec ideal, $V' = \eta_d \dot{m} g_c v$ cu ft/sec actual.

Stagnation State

If a flowing compressible fluid is brought to rest by a reversible process with no shaft work being done, the resulting state is called the *stagnation state.* If the fluid is compressible, the process in isentropic. Let the subscript o designate the stagnation state. $Q = 0$, $W = 0$, and $K_o = 0$. Consider that the fluid flow originates in a large reservoir where $\mathcal{V} \approx 0$. The stagnation enthalpy is

$$h_o = h + K \qquad [\text{AT } s = C]$$

A thermocouple probe in the gas stream tends to measure the stagnation temperature. The temperature of the stream is computed from the isentropic relations. For example, in case of an ideal gas, $\Delta h = c_p \Delta T$. Hence (c_p BTU/lbm-°R), for a small change ΔT,

$$T = T_o - \frac{K}{c_p} = T_o - \frac{\mathcal{V}^2}{2g_c J c_p}$$

$$T_o = T + \frac{K}{c_p}$$

The pressure of the fluid after it is brought to rest with $s = C$ is the stagnation pressure. A pitot tube measures approximately the *stagnation pressure* p_o, which is also called *impact pressure* and *total pressure.* Equation (h) with p_o instead of p_1 is

$$\text{(j)} \qquad \mathcal{V}_2^2 = 2g_c v(p_o - p_2) = 2g_c \left(\frac{p_o - p_2}{\rho}\right) = 2g_c R T_o \left(\frac{p_o - p_2}{p_o}\right)$$

This needs no correction factor for initial velocity, and the rate of flow is computed from (i) where $v_2 \approx v_1$ cu ft/lbm in many flow-measuring instruments and $\mathcal{V}_2$ ft/sec is the velocity of flow at a section whose area is A_2 ft^2.

EXAMPLE 1: What is the flow through a venturi meter with a throat 2 inches in diameter and with an entrance diameter of 4 inches, if the difference in level of an attached mercury manometer is 25 inches? Let the coefficient of discharge be $\eta_d = 98\%$, and assume that the venturi tube is horizontal and that the connecting lines to the manometer are filled with water.

SOLUTION: The problem does not say what is flowing but the fact that the "connecting lines" are filled with water implies strongly that the fluid is water. Moreover, the units in which the flow is to be given are not stated. For liquids, the flow is usually in cu ft/sec, but lbm/sec would probably be accepted. Compute both. If the fluid is water, the 25 inches Hg is partly balanced by 25 inches H_2O. (See figure 10.15.) For water, $62.4/1728 = 0.0361$ psi/inch H_2O.

Then,

$$p_1 - p_2 = (25)(0.49 - 0.0361) = 11.35 \text{ psi}$$

Using (h) and the correction factor I, we have

$$\mathcal{V}_2 = I\sqrt{\frac{2g(p_1 - p_2)}{\rho}} = \sqrt{\frac{2g(p_1 - p_2)}{[1 - (D_2/D_1)^4]\rho}}$$

$$= \sqrt{\frac{(2)(32.2)(11.35)(144)}{[1 - (0.5)^4](62.4)}} = 42.4 \text{ ft/sec}$$

For $A_2 = \pi D^2/4 = \pi \text{ in}^2$,

$$V = 42.4\pi/144 = 0.925 \text{ cu ft/sec}$$

$$w = (62.4)(0.925) = 57.6 \text{ lbm/sec, the ideal flow}$$

The actual flow $\dot{m}'g_c = \eta_d \dot{m} g_c = (0.98)(57.6) = 56.5$ lbm/sec. (*Ans.*) These results evidently ignore the effects of temperature.

EXAMPLE 2: A $3 \times 1\frac{1}{2}$ venturi, with a coefficient of discharge of 98%, is used to measure the flow of air in a 3-inch ID pipe. Barometer = 30.05 inches Hg; temperature of the air in the pipe = 114 °F and its static pressure is $p_1 = 0.287$ inches Hg gauge; pressure drop in venturi = 0.838 inch Hg. What quantity of air (lbm/sec) is flowing?

SOLUTION: Convert units: $p_1 = (0.287)(0.49) = 0.1408$ psig. For $(30.05)(0.49) = 14.74$ psia, $p_1 = 14.74 + 0.1408 = 14.88$ psia.

$$v_o \approx v_2 \approx v_1 = \frac{RT_1}{p_1} = \frac{(53.3)(114 + 460)}{(14.88)(144)} = 14.27 \text{ cu ft/lbm}$$

Using equation (j) above and the initial velocity correction factor I, we find (pressure unit cancels):

$$\mathcal{V}_2 = \sqrt{\frac{2gRT_1}{1-(D_2/D_1)^4}\left(\frac{p_1-p_2}{p_1}\right)}$$

$$= \sqrt{\frac{(2)(32.2)(53.3)(574)(0.838)}{(1-0.5^4)(30.05+0.287)}} = 241 \text{ ft/sec}$$

$$\dot{m}' g_c = \eta_d \frac{A_2 \mathcal{V}_2}{v_2} = \frac{(0.98)(\pi 1.5^2)(241)}{(4)(144)(14.27)} = 0.203 \text{ lbm/sec} \quad (Ans.)$$

OPEN CHANNEL FLOW

A semicircular open channel, more expensive to build than rectangular or trapezoidal, will discharge more water than any other shape of the same area of liquid cross section A, given the same slope S and roughness coefficient n. This is because the wetted perimeter is a minimum, the hydraulic radius R is a maximum, and, hence, the velocity is a maximum.

$$\mathcal{V} = C\sqrt{RS} \text{ ft/sec} \qquad \text{[CHEZY EQUATION]}$$

$$C = \left(\frac{1.49}{n}\right)\sqrt[6]{R} \qquad \text{[MANNING FORMULA]}$$

Using both of these expressions in $V = A\mathcal{V}$ cu ft/sec (a common symbol for quantity in cu ft/sec is Q), we get

$$\text{(k)} \qquad V = \left(\frac{1.49}{n}\right) AR^{2/3}\sqrt{S} \text{ cu ft/sec}$$

$S = \tan\theta$, where θ is the angle that the bottom surface of the channel makes with the horizontal.

Typical values of n are: smooth wood, 0.012; rough concrete, 0.014; rubble masonry, 0.017; gravel, 0.020; and canals, (and rivers with rough bottoms and plant growth), 0.035.

To use minimum materials for a rectangular or trapezoidal channel, let $R = d/2$, where d = depth of flow. In a rectangular channel, the corresponding shape turns out to be

$$\text{depth of flow} = \frac{1}{2}\text{ (width of channel)}$$

EXAMPLE: A circular sewer on a slope of 4 EE–4 has a value of Manning's $n = 0.013$. It carries 25 cu ft/sec when flowing half full. Find its diameter.

SOLUTION: Semicircular area $= \pi r^2/2$.

$$R = \frac{\pi r^2}{2 \times \pi r} = \frac{r}{2}$$

r = radius of sewer. Use these values in the Chezy-Manning equation (k):

$$V = \frac{1.49 A R^{2/3} \sqrt{S}}{n}$$

$$25 = \frac{(1.49)(\pi r^2) r^{2/3} (0.0004^{0.5})}{(0.013)(2) 2^{2/3}}$$

From this, $r = 2.46$ feet, and $D = 4.92$ feet. (*Ans.*) If this computation is for the purpose of selecting sewer size, a nominal 5-foot size would be specified.

FLOW IN PIPES

Viscosity

Shearing stress on fluid (F/A) is proportional to the rate of shearing strain (or velocity gradient) $d\mathcal{V}/dy$ and the proportionality constant μ is called the ***absolute*** or ***dynamic viscosity.***

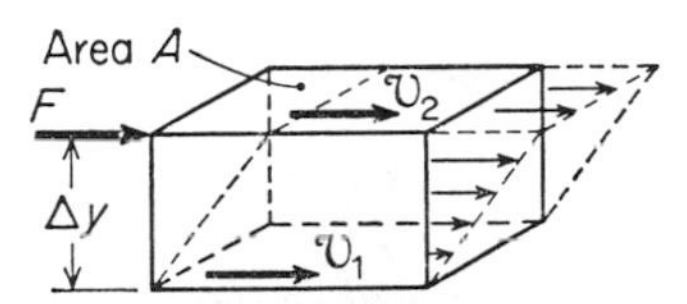

Figure 10.3

(1) $$\frac{F}{A} = \mu \frac{\Delta \mathcal{V}}{\Delta y}$$ [FIGURE 10.3]

Note that the units of μ include the unit of force (also, $\Delta \mathcal{V} = \mathcal{V}_2 - \mathcal{V}_1$, as in figure 10.3). See *Consistent Units* below. Metric units are:

$$\mu \text{ poise} = \frac{\text{dyne-sec}}{\text{cm}^2}$$

Conversion constants are:

$$100 \, \frac{\text{centipoise}}{\text{poise}}$$

$$1490 \, \frac{\text{centipoise}}{\text{lbm/sec-ft}} \qquad \text{[MASS IN LBM]}$$

$$47{,}900 \, \frac{\text{centipoise}}{\text{lbf sec/ft}^2} \qquad \text{[FORCE IN LBF]}$$

Kinematic viscosity $\nu = \mu/\rho$ usually is given in stokes or centistokes (100 centistokes/stoke), μ and ρ in consistent units.

Metric units for ν are cm^2/sec = stokes. English units are ft^2/sec.

To convert to English units, convert cm^2 to ft^2. The viscosity of liquids generally decreases as the temperature rises. Viscosity of gases increases as the temperature increases. Viscosity changes very little with pressure, except for extreme changes of pressure. Values of viscosities are given in handbooks and texts.

Consistent Units

A system of consistent units is based on Newton's law in the form $F = ma$, in which mass is defined in terms of force and force in terms of mass. If force is taken in pounds, the unit of mass is derived from

$$m = \frac{F}{a} \text{ in units of } \frac{\text{lbf}}{\text{ft/sec}^2} = \frac{\text{lbf-sec}^2}{\text{ft}} = \text{slug} \qquad [a \text{ ft/sec}^2]$$

If mass is taken in pounds, the unit of force is derived from

$$F = ma \text{ in units of lbf-ft/sec}^2 = \text{poundal}$$

When it is said that units must be consistent, one must use pounds *only for force* or *only for mass*, deriving the other unit from $F = ma$.

Friction Factor

The *head loss* or *friction head*, h_f, of a fluid flowing full in a round pipe is given by the Darcy equation:

$$h_f = f \frac{L \mathcal{V}^2}{D 2g} \text{ ft} \qquad \text{(m)}$$

Units might also be in ft-lbf/lbm of mass. This equation works whether the flow is streamline or turbulent. L ft is the length of pipe, D ft is its ID, $\mathcal{V}$ ft/sec is the average fluid speed, and f is the pipe friction factor. The friction factor f is a function of the Reynolds number $\mathbf{R}$ and the relative surface roughness ϵ/D, where ϵ represents the effective or relative roughness. Values of ϵ and ϵ/D have been determined for various commercial pipes and are found in handbooks. The pressure drop corresponding to h_f is

$$\Delta p = (h_f \text{ ft})(\gamma \text{ lbf/cu ft}) \text{ lbf/ft}^2$$

The Reynolds number is defined by

$$\text{(n)} \qquad \mathbf{R} = \frac{D\mathcal{V}\rho}{\mu} = \frac{D\mathcal{V}}{v} = \frac{DG}{\mu} \quad \text{(dimensionless)}$$

The units must be consistent: ρ = density (lbm/cu ft for pound mass, slugs/cu ft for pound force); μ = absolute viscosity (lbm/sec-ft for pound mass, lbf-sec/ft^2 for pound force); $\nu = \mu/\rho$ = kinematic viscosity (ft^2/sec); $G = \mathcal{V}\rho$ = mass rate of flow per unit area.

The friction factor f is obtained from charts such as figure 10.4, or it is computed from equations:

$$f = \frac{64}{\mathbf{R}}$$ [DARCY; LAMINAR FLOW, *AB* IN FIGURE 10.4]

$$f = \frac{0.316}{\mathbf{R}^{0.25}}$$ [BLASIUS; TURBULENT FLOW IN SMOOTH PIPES, 3000 < R < 10,000]

$$\frac{1}{\sqrt{f}} = 1.74 - 2\log\frac{18.6}{\mathbf{R}\sqrt{f}}$$ [VON KARMAN; SMOOTH PIPE FLOW]

$$\frac{1}{\sqrt{f}} = -2\log\left(\frac{2.51}{\mathbf{R}\sqrt{f}} + \frac{\epsilon}{3.7D}\right)$$ [COLEBROOK; TRANSITION ZONE IN FIGURE 10.4]

Except under certain laboratory conditions, laminar flow ceases at about $\mathbf{R}$ = 2100, then enters a critical zone (2100 to 3000), a transition zone, and, finally, complete turbulence with roughness projections beyond the laminar boundary layer. The resistances of elbows, valves, and so forth are found in handbooks, usually in feet of friction head.

EXAMPLE: Calculate the pressure drop in psi per 1000 feet of 2-inch ID pipe used to carry 100 gpm of linseed oil at 70 °F if the flow is laminar. The absolute viscosity of the oil is 0.0014 lbf-sec/ft^2. Its specific gravity is

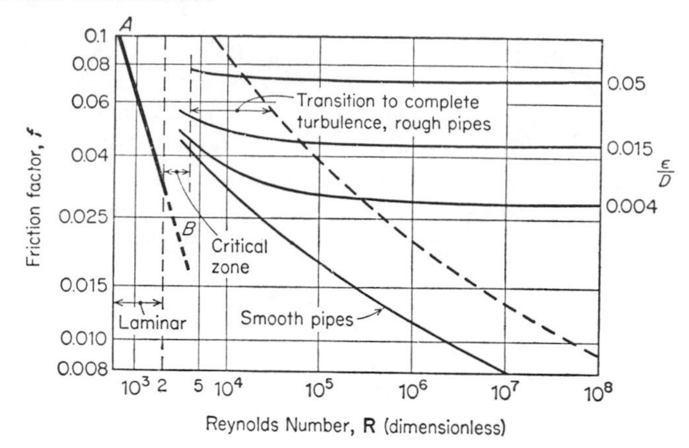

Figure 10.4 Friction factors for pipe flow.
(After L. F. Moody, *ASME Trans.*, **66**, p. 671.)

0.90. The average velocity in this pipe is one-half the centerline velocity. The critical Reynolds number for laminar flow is to be taken as 2100.

SOLUTION: For a pipe area of $A = \pi D^2/4 = \pi/144$ ft² and a flow of $V = (100 \text{ gpm}/7.48) = 13.36$ cu ft/min., the average speed of flow is

$$\mathcal{V} = \frac{V}{A} = \left(\frac{13.36}{60}\right)\left(\frac{144}{\pi}\right) = 10.2 \text{ ft/sec}$$

Comparing the given units of μ (lbf-sec/ft²) with the above basic equation (1) defining viscosity, we see that the pound-force system is used. Hence, the density in the equation for **R** must be in slug units to match. (See "Consistent Units," above.)

$$\begin{aligned} \rho &= (\text{specific gravity})(\text{water density}) \\ &= (0.9)(62.4/32.2) = 1.745 \text{ slugs/cu ft} \end{aligned}$$

$$\mathbf{R} = \frac{D\mathcal{V}\rho}{\mu} = \frac{(2/12)(10.2)(1.745)}{0.0014} = 2120$$

This is close to the specified dividing line of 2100, and the problem definitely says *laminar flow.* It also has been assumed that the specific gravity as given is for the 70 °F temperature of the oil. Thus, for a friction factor of $f = 64/\mathbf{R} = 64/2120$, the friction head loss becomes

$$h_f = \frac{fL\mathcal{V}^2}{D2g} = \frac{(64)(1000)(10.2)^2}{(2120)(2/12)(2)(32.2)} = 291 \text{ ft}$$

Since the implication is that terrestrial gravity is closely standard, the specific weight is the same as the density: $\gamma = (0.9)(62.4) = 56.16$ lbf/cu ft and

$$p_1 - p_2 = \frac{(56.16)(291)}{144} = 113.5 \text{ psi} \quad (Ans.)$$

Power to Maintain Flow

The power required to pump a fluid (liquid or uncompressed gas) is proportional to the weight of fluid W times the total head, h_t feet of the fluid being pumped; $W = V\gamma$. With the time unit of second, W is in lbf/sec, V is in cu ft/sec (550 ft-lbf/hp-sec), and work $= Wh_t = V\gamma h_t$ ft-lbf/sec.

$$\text{hp} = \frac{Wh_t}{550} = \frac{V\gamma h_t}{550} \quad \text{(n)}$$

h_t is the sum of all the heads, from equation (f) (pressure head, velocity head, gravitational head, and friction head). If the efficiency of the pump is η, the actual input power to the pump must be

$$\text{hp from (n)}/\eta$$

MISCELLANEOUS

Speed of Sound in a Fluid

The speed of sound in a fluid is Mach 1, given by

$$a = \underset{(1)}{\sqrt{g_c kRT}} = \underset{(2)}{\sqrt{g_c Ev}} = \underset{(3)}{\sqrt{\frac{g_c kp}{\rho}}} = \underset{(4)}{\sqrt{\frac{g_c E}{\rho}}} \text{ ft/sec} \quad \text{(o)}$$

$k = c_p/c_v$, $R =$ gas constant, ft-lbf/lbm-°R, T is in °R, $g_o = 32.2$ ft/sec^2, E lbm/ft^2 = *bulk modulus* (also called *elasticity of fluid*, experimentally determined), v cu ft/lbm, p lbf/ft^2, ρ lbm/cu ft. (Note the mixed system of units. In a consistent system, $a = \sqrt{E/\rho}$.) If water has a bulk modulus of $E = 300{,}000$ psi (at about 45 °F) and a density of 62.4 lbm/cu ft, the velocity of sound in it is, from (4) in equation (o) above,

$$a = \sqrt{(32.2)(144 \times 300{,}000)/62.4} = 4720 \text{ ft/sec}$$

For a gas, use (1) or (2). Thus, in 70 °F atmospheric air,

$$a = \sqrt{(32.2)(1.4)(53.3)(530)} = 1128 \text{ ft/sec}$$

With few exceptions, notably lead, the speed of sound in metals is much greater than that in either liquids or gases.

Drag

A body, such as an airfoil, moving through a fluid encounters a resistance to motion called the *drag* in the amount of

$$D = C_D A \frac{\rho \mathcal{V}^2}{2g_c} \text{ lbf}$$

C_D is an experimentally determined factor (dimensionless), A ft^2 is a significant area (the area of the projection of the airfoil on the plane of the chord), $\mathcal{V}$ ft/sec is speed, and ρ is the density in lbm/ft^3.

Force at a Pipe Bend

Use the principle of impulse and momentum. In the vicinity of the bend, the speed $\mathcal{V}$ is virtually constant. $F = (\dot{m}\Delta\mathcal{V})$ where (V cu ft/sec)(ρ lbm/cu ft) is the mass rate of flow. In figure 10.5, summing in x and y directions,

$$-F_x = \dot{m}(\mathcal{V}\cos\theta - \mathcal{V})$$

$$F_y = \dot{m}(\mathcal{V}\sin\theta - 0)$$

$$F = \sqrt{F_x^2 + F_y^2}$$

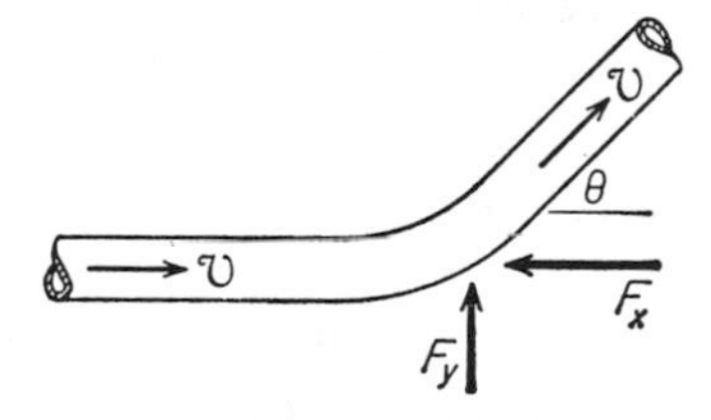

Figure 10.5

PRACTICE PROBLEMS

10.1. The barometer reads 29.0 inches Hg. Calculate the absolute pressure in a tank whose attached gauge reads 9.5 psi vacuum.

Ans. 4.7 psia

10.2. What column of water will balance the pressure exerted by 760 millimeters of mercury? What column of gasoline (specific gravity = 0.67)?

Ans. 33.75 ft, 50.4 ft

10.3. A piece of lead (specific gravity = 11.3) is tied to 8 cubic inches of cork whose specific gravity is 0.25. They float submerged in water. What is the weight of the lead?

Ans. 0.238 lbf

10.4. (a) Find the static pressure (in feet) of water at the center of the pipe shown in figure 10.6. (b) Find the velocity of the water flowing in the pipe.

Ans. (a) 2.27 ft H_2O gauge, (b) 15.11 ft/sec

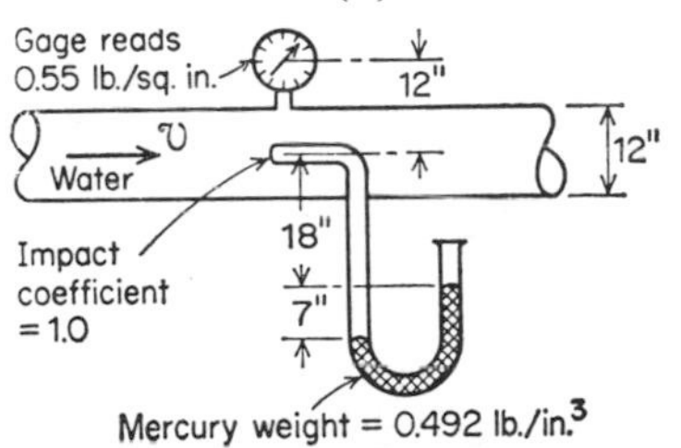

Figure 10.6

10.5. Calculate the weight per second of water flowing past a point in an 8-inch pipe if the velocity is 12.0 ft/sec.

Ans. 261.5 lbf/sec

10.6. 1000 gpm of water flows through a line of varying diameter with one end (point 1) being 10 feet lower than the other end (point 2). If the diameter at point 1 is 8 inches and at point 2 is 4 inches, what is the static pressure (psi) at point 1 if the pressure at point 2 is atmospheric? Assume flow to be frictionless.

Ans. 8.45 psig

10.7. A 3-inch hose discharges through a nozzle having an opening of 2 inches diameter. If the hose is attached to a tank 100 feet below the surface of the water, what will be the discharge in gallons per minute? Assume no friction in the tank or hose and that c_v is 0.98. What will be the height to which the jet will rise if the nozzle is held vertically at the elevation of the outlet of the tank?

Ans. 770 gpm, 96 ft

10.8. A fire hose has a 3-inch inside diameter and is fitted with a conical nozzle 2 inches ID at the outlet. Find the quantity discharged through the nozzle if the pressure at the base of the nozzle is 100 psig and the coefficient of discharge is 0.98.

Ans. 2.91 cu ft/sec

10.9. A pipeline has a gradual change of diameter from 12 inches to 6 inches. The small end is 15 feet lower than the large end. If the static pressure at both ends is the same, what is the flow in cu ft/sec?

Ans. 6.31 cu ft/sec

10.10. What total stress exists in cable AB of the crane in figure 10.7 while holding a pontoon above the water? The pontoon is a cylinder 4 feet in diameter and 60 feet long, outside dimensions, and will float half submerged.

Ans. 23,600 lbf

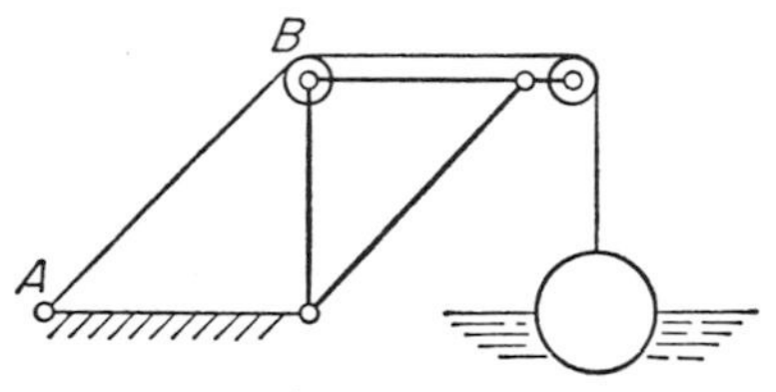

Figure 10.7

10.11. A water flume is dammed by a rectangular gate 10 feet wide and 6 feet high. The water rises to within 1 foot of the upper edge of the gate. (a) Calculate the total water pressure force against the upstream side of the gate, if the downstream portion of the flume is empty. (b) How far below the upper edge of the gate does the resultant of the total water pressure force act?

Ans. (a) 7800 lbf, (b) 4.33 ft

10.12. A completely enclosed tank with a base 4 ft $\times$ 4 ft has a pipe protruding from the top. (See figure 10.8.) Water fills the tank and the pipe to a depth of 7 feet from the bottom. (a) What is the total pressure on a side of the tank? (b) What is the total pressure on the bottom of the tank?

Ans. (a) 4120 lbf, (b) 7000 lbf

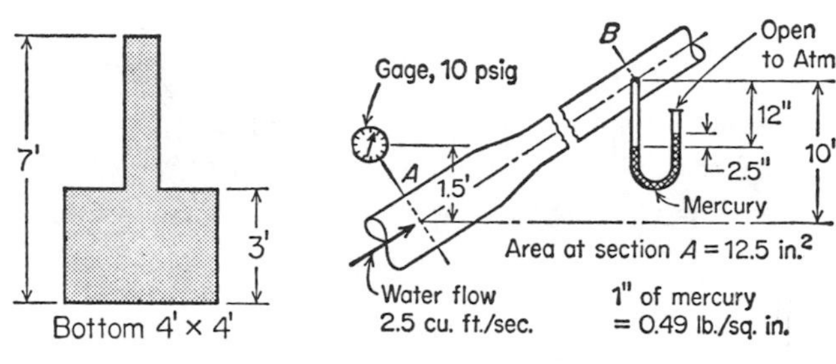

Figure 10.8 **Figure 10.9**

10.13. From the information shown on the schematic diagram in figure 10.9, calculate the velocity of the fluid at section B in ft/sec.

Ans. 40.6 ft/sec

10.14. In a centrifugal pump test the discharge gauge reads 100 psi and the gauge on the suction reads 5 psi. Both gauges indicate pressure above atmospheric. The gauge centers are at the same level. The diameter of the suction pipe is 3 inches, and the diameter of the discharge pipe is 2 inches. Suction and discharge are at the same level. Oil (specific gravity = 0.85) is being pumped at the rate of 100 gpm. Assuming no friction losses, what energy in foot-pounds is added by the pump to each pound of oil that passes through it?

Ans. 260

10.15. A water-tight cubical box 12 inches outside dimensions is made of iron plates $\frac{1}{4}$ inch thick, the iron weighing 480 lbm/cu ft. Will the box float in water? Why?

Ans. Yes, weight of box is 57.5 lbf

10.16. A hydraulic jack has a ram diameter of $1\frac{1}{2}$ inches and a pump piston diameter of $\frac{5}{8}$ inch. The jack handle has a mechanical advantage of 15 to 1. (a) Neglecting friction, what weight will the jack lift when 30 lbf is applied to the handle? (b) What is the pressure of the hydraulic fluid under these conditions?

Ans. (a) 2592 lbf, (b) 1465 psi

10.17. Brine (specific gravity = 1.20) flows through a pump at 2000 gpm. The pump inlet is 12 inches in diameter. At the inlet, the vacuum is 6 inches of mercury. The pump outlet, 8 inches in diameter, is 4 feet above the inlet. The outlet pressure is 20 psig. What power does the pump add to the fluid?

Ans. 30.3 hp

10.18. A water pump develops a total head of 200 feet. The pump efficiency is 80% and the motor efficiency is 87.5%. If the power rate is \$0.015 per kw-hr, what is the power cost for pumping 1000 gallons?

Ans. \$0.0135

10.19. A pump shown in figure 10.10 pumps water to an elevation of x feet above the center of the pump. When the pump is running, the pressure gauge A registers 39.2 psi. When the pump is shut off, the pressure gauge A reads 10 psi due to the water in the line. A characteristic curve, as furnished by the manufacturer for the pump, is shown. (a) What is the total head, including the suction head, against which the pump must operate if the friction in the inlet line can be considered negligible? (b) What is the head loss due to friction in the line from the gauge to discharge? (c) How many gallons per minute are being pumped when the pressure gauge registers 39.2 psig? (d) Calculate the efficiency of the pump with 1.25 hp input and 39.2 psig discharge pressure.

Ans. (a) 100 ft, (b) 67.4 ft, (c) 30 gpm, (d) 60.5%

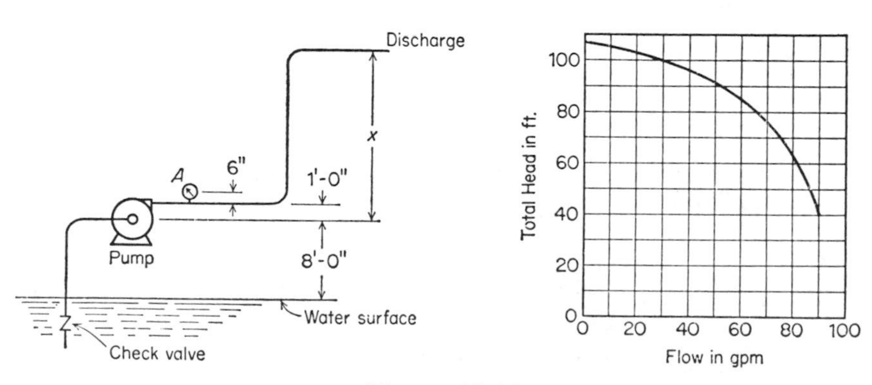

Figure 10.10

10.20. Two horizontal smooth pipes of equal length carry air and water respectively such that the Reynolds numbers and pressure drops for each are the same. Find the ratio for the mean air velocity to that of the water.

Ans. $(\nu_{air}/\nu_{water})^{1/3}$

10.21. Figure 10.11 shows a portion of a pipeline in which water is flowing at a rate of 0.87 cu ft/sec. At section A, ID = 4.026 inches and cross-sectional area is 0.0885 ft^2; at B, ID = 3.068 inches and area = 0.0514 ft^2. (a) What will the pressure gauge at B read if the head loss due to pipe friction between A and B is 7.50 ft? (b) What is the Reynolds number at B if the kinematic viscosity is 1.08 EE–5 ft^2/sec?

Ans. (a) 26.8 psi, (b) 400,000

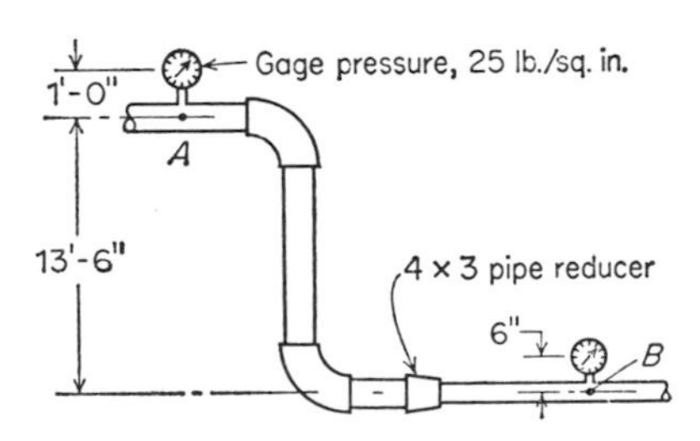

Figure 10.11

10.22. In figure 10.12, flow at A = 1.25 cu ft/sec. The actual inside diameter of pipe is 4.02 inches. The actual inside area of pipe is 12.7 in^2. The friction factor for 4-inch pipe is 0.024. Determine the following: (a) the total loss of head in feet from A to B due to fluid friction in the pipe and fittings, (b) the head loss due to pipe friction, (c) the head loss due to valves and fittings.

Ans. (a) 147.4 ft H_2O, (b) 134 ft H_2O, (c) 13.4 ft H_2O

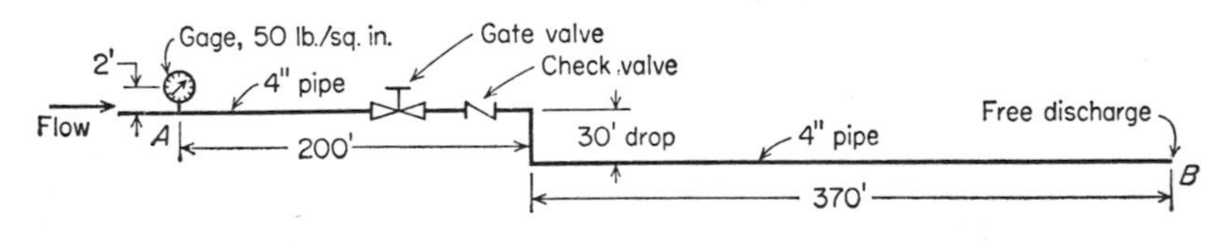

Figure 10.12

10.23. A centrifugal motor-driven water pump is connected to a pipeline as shown in figure 10.13. Pipe ID = 2.067 inches, pipe flow area = 3.36

in^2 (7.48 gal/cu ft). (a) What is the average velocity in the pipe? (b) If the input to the pump motor is 1105 watts, what is the combined efficiency of the motor and pump? (c) What is the friction factor for the 300-foot length of pipe?

Ans. (a) 3.82 ft/sec. (b) 42.2%, (c) 0.030

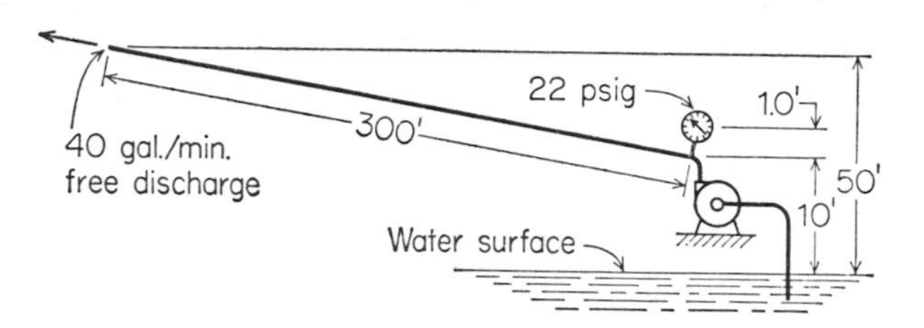

Figure 10.13

10.24. A jet of water is discharged through a 1-inch diameter orifice under a head of 2.10 feet. The total discharge is 228 lbm in 90 seconds. The jet is observed to pass through a point 2 feet downward and 4 feet away from the vena contracta. See figure 10.14. Compute the coefficient of contraction and the coefficient of velocity.

Ans. $C_c = 0.655$, $C_v = 0.978$

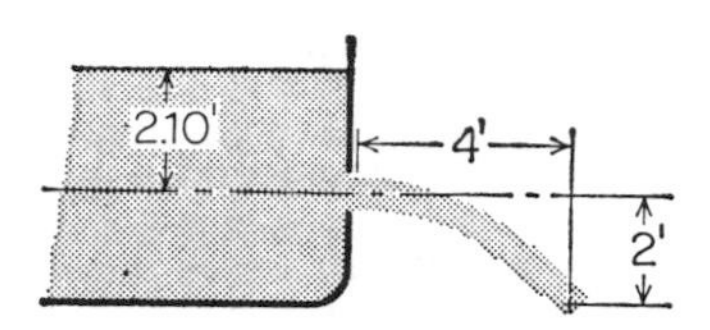

Figure 10.14

10.25. A 2-inch jet of water from a nozzle is pointed upward at 60° to the horizontal. The jet reaches a height of 215 feet above the ground. (a) Find the velocity of the jet as it leaves the nozzle. (b) What horsepower is available in the jet at the nozzle? (c) What quantity of water is discharged, in cu ft/sec?

Ans. (a) 136 ft/sec, (b) 96.5 hp, (c) 2.96 cu ft/sec

10.26. A penstock 36 inches in diameter carries 138 cu ft/sec water. (a) Find the thrust on a 20° elbow. (b) Find the force perpendicular to the original direction of flow.

Ans. (a) $F_x = 321$ lbf, $R = 1820$ lbf, (b) $F_y = 1785$ lbf

10.27. A venturi meter with an 8-inch diameter throat is installed in a 12-inch diameter water pipeline as shown in figure 10.15. When the mercury manometer registers a differential of 4 inches, what is the flow in cu ft/sec? Assume the venturi coefficient to be 1.00, and the specific gravity of the mercury to be 13.6.

Ans. 6.4 cu ft/sec

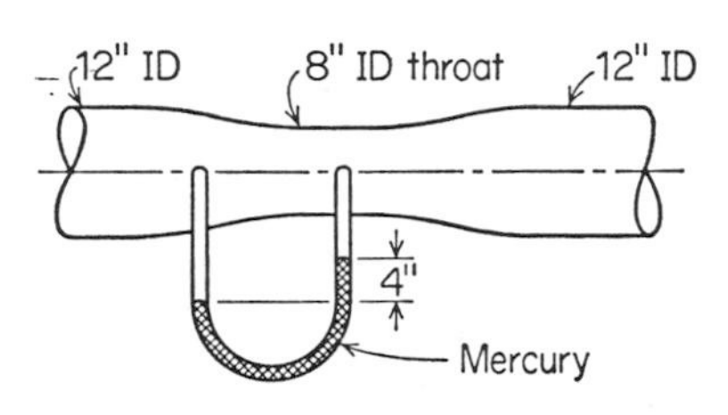

Figure 10.15

10.28. What slope is necessary for a flume whose cross section is shown in figure 10.16, in order that a flow of 4.0 cu ft/sec can be maintained? Assume a Manning's friction factor of $n = 0.012$.

Ans. 0.0032

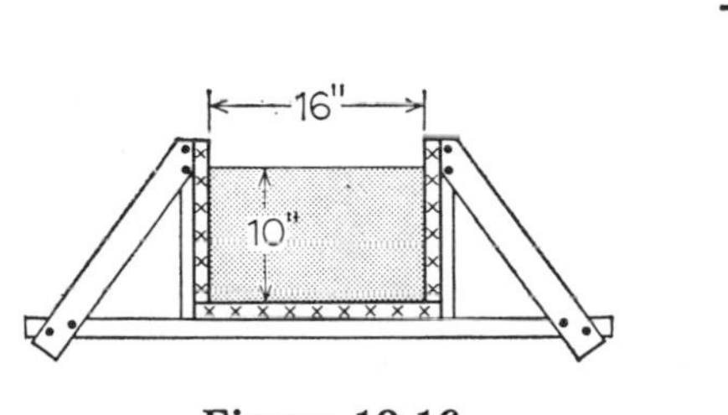

Figure 10.16

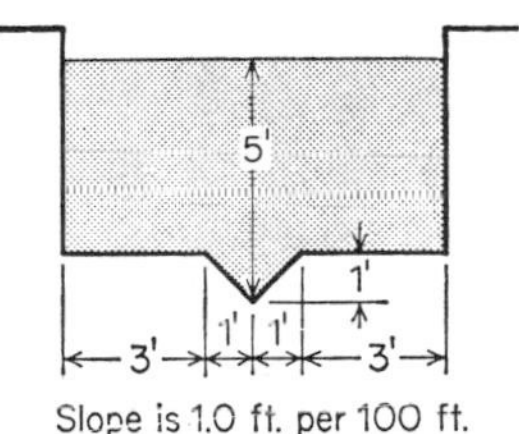

Figure 10.17

10.29. (a) Find the hydraulic radius of the section shown in figure 10.17 when the water is flowing at a depth of 5.0 feet from the bottom of the V-notch. (b) What quantity of water in cu ft/sec will flow through this section, at the above-mentioned depth, if the coefficient of roughness is 0.013?

Ans. (a) 1.96 ft, (b) 590 cu ft/sec

10.30. Calculate the dimensions of a flume with the smallest wetted perimeter possible that will carry a flow of 50 cu ft/sec at a slope of 1 foot per 1000 feet of length. Assume a friction coefficient of $n = 0.014$.

Ans. semicircular, $r = 2.77$ ft

10.31. Let the axis of the pipe shown in figure 10.18 be horizontal, the diameter at A being 24 inches and the diameter at B, 12 inches. Assume the rate of discharge to be 8 cu ft/sec in the direction AB, and the pressure head at the center of the section A to be 20 feet. Compute the resultant force exerted upon the pipe by this portion of the stream. Assume there is no head loss.

Ans. 2896 lbf

Figure 10.18

10.32. A 1-inch diameter cable is used to anchor a buoy near the entrance to a harbor where the ocean current reaches a velocity of 4 mph. A model is made of the buoy and cable to 1/10 scale for tests in the laboratory. Using fresh water in the test, results show that for a velocity of 5 ft/sec, the drag on 1 foot of model cable is 2 lbf. The density of sea water is 63.8 lbm/cu ft. The relation between the model and prototype can be expressed as $F/(\rho AV^2)$ = a constant, where F is drag force, ρ is density of fluid, A is area (projected area perpendicular to flow), and V is velocity of fluid relative to model or prototype. What is the drag per foot of prototype cable as installed in the harbor entrance?

Ans. 28.2 lbf/ft

10.33. If the pressure of a wind blowing 20 mph is 800 lbf on a wall of 1000 square feet area, what would be the pressure of a 40 mph wind on the same wall?

Ans. 3200 lbf

10.34. Sound travels fastest in (a) air at 70 °F and 1 atm pressure, (b) air at 0 °F and 1 atm pressure, (c) hydrogen at 0 °F and 1 atm pressure, (d) water at 70 °F, (e) steel at 70 °F.

Ans. (e)

10.35. How much oil in gpm is flowing through a venturi meter with a 4-inch throat placed in a pipeline 6 inches in diameter? A mercury-oil differential gauge connected between the upstream section and the throat in such a manner as to record the difference in pressure heads shows a difference of 8 inches. The specific gravity of the oil is 0.95, and the discharge coefficient is 1.00. Specific gravity of mercury is 13.6.

Ans. 1048 gpm

10.36. Find the hydraulic radius of the section shown in figure 10.19 when the water is flowing 4 feet deep in the channel.

Ans. 2.58 ft

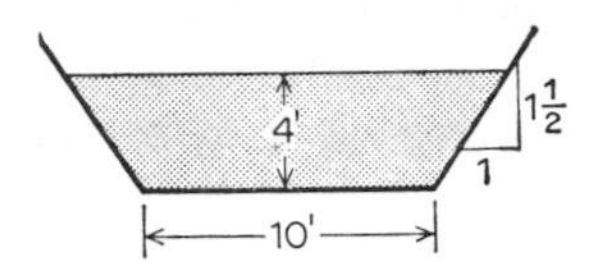

Figure 10.19

10.37. The Mach number is defined as the ratio of velocity to the speed of sound. At what Mach number is an airplane flying if it travels at a true airspeed of 1200 mph over the desert (assume dry air) where atmospheric temperature is 100 °F?

Ans. 1.52

10.38. At point A in a certain pipe that carries water, the diameter is 6 inches and the pressure is 10 psig, while at point B, 10 feet below point A, the diameter is 12 inches and the pressure is 15 psig. Determine the direction of flow and the friction loss between the two points when the discharge is 1.0 cu ft/sec.

Ans. B to A, 1.12 ft

10.39. A tank separated in the center by a vertical partition contains water on one side to a depth of 8 feet. The other side contains another fluid (whose specific gravity is 1.30) to a depth of 10 feet. A rectangular opening in the center partition 2 feet wide by 3 feet high is closed by a flat plate. The plate is hinged at its upper edge which is 3 feet above the bottom of the tank. What force applied perpendicular to the plate at its lower edge is necessary to keep it closed?

Ans. 880 lbf

10.40. Points A and B are 3000 feet apart along a new 6-inch inside diameter steel pipe. B is 60 feet above A in elevation. There are flowing in this pipe 750 gpm of a fuel oil (specific gravity = 0.90) whose dynamic viscosity is 0.0015 lbf-sec/ft^2. The direction of flow is from A to B. The friction factor for this pipe is $f = 0.03$. What pressure must be maintained at A if the pressure at B is to be 50 psi?

Ans. 153 psi

10.41. A 3-inch diameter pipe, 2000 feet long with a friction factor f of 0.020 carries water from a reservoir and discharges freely at a point

100 feet below the reservoir level. Find the pump horsepower required to double the gravity flow.

Ans. 21.1

10.42. A 4-foot $\times$ 6-foot rectangular plate is immersed vertically in water with one of its 4-foot sides in the water surface. A straight line is drawn from one end of the 4-foot side in the surface ($x = 0$) to a point on the other 4-foot side so as to divide the rectangle into two areas. Where must this line intersect the other 4-foot side to make the total forces of water pressure on the two areas equal?

Ans. $x = 3$

10.43. A cylindrical tank 4 feet in diameter has its axis horizontal. At the middle of the tank, on top, is a pipe 2 inches in diameter, which extends vertically upward. The tank and pipe are filled with an oil (specific gravity $= 0.85$) with the free surface in the 2-inch pipe at a level of 10 feet above the tank top. What is the total force on one end of the tank?

Ans. 8020 lbf

10.44. (a) What is the theoretical horsepower of a hydroelectric plant that uses 1200 cu ft/sec of water under a head of 150 feet? (b) If the efficiencies of the approaches, the turbines, and the dynamos are 99%, 76%, and 96%, respectively, what power in kilowatts is delivered?

Ans. (a) 20,400 hp, (b) 11,000 kw

10.45. A gravity dam composed of concrete is 45 feet high. It is 3 feet wide on top and the base is 30 feet wide. The water side of the dam is vertical and the other side slopes uniformly from top to bottom. The water behind the dam is 42 feet deep and there is no water pressure under the dam. Compute: (a) the position of the resultant at the base of the dam, (b) the pressure (in lbf/ft^2) at the toe and at the heel of the dam, (c) the position of the resultant with the reservoir empty. (d) At what minimum friction factor would the dam be safe against sliding?

Ans. (a) 17 ft from H_2O face, (b) 5200 lbf/ft^2, 2220 lbf/ft^2, (c) 10.1 ft, (d) $f = 0.495$

10.46. A cylindrical tank is 20 feet in diameter and 40 feet high. How long will it take to lower the water from the 40-foot level to the 20-foot level through a 4-inch orifice at the bottom having a discharge coefficient of 0.98?

Ans. 1690 seconds

10.47. (a) A smooth nozzle on a fire-hose is $1\frac{1}{8}$ inches in diameter and discharges 250 gpm. What is the force with which this stream of water strikes a flat surface normal to the stream if placed only a short distance away? (b) What is the velocity head of the jet?

Ans. (a) 87 lbf, (b) 101 ft

10.48. A pipe, 12 inches in diameter, carries water by gravity from a reservoir. At a point 500 feet from the reservoir, and 28 feet below its surface, a pressure gauge reads 10.5 psi. At a point 8500 feet from the reservoir and 280.5 feet below its surface, a pressure gauge reads 61.5 psi. Compute the discharge.

Ans. 3050 gpm

10.49. Water issues from a circular orifice 4 inches in diameter. The height of water above center of orifice is 30 feet. Assuming coefficient of discharge of 0.72, what is the discharge?

Ans. 165.5 cu ft/min

10.50. Water flows through 3000 feet of 36-inch diameter pipe which branches into 2000 feet of 18-inch diameter pipe and 2400 feet of 24-inch diameter pipe. These rejoin and the water continues through 1500 feet of 30-inch diameter pipe. All pipes are horizontal and the friction factors are: 0.016 for the 36-inch pipe, 0.017 for the 24-inch and 30-inch pipe, and 0.019 for the 18-inch pipe. Find the *pressure drop* between the beginning and the end of the system, if the steady flow is 60 cu ft/sec in the 36-inch pipe. Neglect minor losses.

Ans. 40 psi

10.51. A horizontal bend in a pipeline reduces the pipe from a 30-inch diameter to an 18-inch diameter while bending through an angle of 135° from its original direction. The flow rate is 10,000 gpm and the direction of flow is from the 30-inch to the 18-inch pipe. If the pressure at the entrance is 60 psi gauge, what must be the magnitude of the resultant force necessary to keep the bend in place? Consider the bend to be adequately supported in a vertical direction.

Ans. 55,000 lbf

10.52. An outfall sewer 48 inches in diameter is 3000 feet long. The invert elevation at the beginning is 6.5 feet higher than the outlet into a stream. (a) If $n = 0.013$, determine both the capacity of the sewer when flowing full and the velocity. At periods of high water the sewer outlet can be submerged so the water level is 5 feet above the sewer invert. (b) What will be the capacity of the sewer when flowing full, and the velocity during high water?

Ans. (a) 67 cu ft/sec, 5.3 ft/sec, (b) 49 cu ft/sec, 3.9 ft/sec

10.53. A nozzle delivers water to a Pelton wheel under 1200-foot head. If the pipe loss is 5% and the nozzle efficiency is 98%, what is the velocity of the jet? With a wheel efficiency of 88%, how much water is required to develop 5000 horsepower?

Ans. 268 ft/sec, 20,100 gpm

10.54. A tank is filled with water. A hole is drilled through the tank 8 feet below the water surface. What is the velocity of the water passing through the hole (in ft/sec)?

Ans. 22.7 C_v ft/sec

10.55. Water stands 10 feet above the bottom of a tank. What is the gauge-pressure reading at the bottom?

Ans. 4.33 psi

10.56. An 8-cubic-foot block of concrete, weighing 150 lbm/cu ft is to be used as an anchor in fresh water. Assuming the pull on the anchor will be at 45° with the vertical, what is the pulling force that the anchor will withstand?

Ans. 990 lbf

11 *HEAT TRANSFER*

SYMBOLS

A = area
c_p = specific heat at constant pressure
D = diameter
F_A = angle factor
F_E = emissivity factor
$\mathbf{h}$ = film coefficient
$k = c_p/c_v$
$\mathbf{k}$ = thermal conductivity
m = mass in slugs
$\mathbf{N}$ = Nusselt number
$\mathbf{P}$ = Prandtl number
Q = heat, BTU
R = thermal resistance
$\mathbf{R}$ = Reynolds number
T = absolute temperature, °R
t = temperature, °F
U = over-all transmittance
$\mathcal{V}$ = speed
ϵ = emissivity
μ = absolute viscosity
ν = kinematic viscosity
ρ = density
σ = Stefan-Boltzmann constant
Δ = a change of, or a difference

CONDUCTION

Plane Wall

Conducted heat is computed by Fourier's law:

$$Q = -\mathbf{k}A\,\frac{dt}{dL}$$

For steady-state flow in one direction only with $\mathbf{k}$ and A constant, the integrated form is

(a) $$Q = -\mathbf{k}A\frac{\Delta t}{L} = \frac{A(-\Delta t)}{L/\mathbf{k}} \text{ BTU/hr}$$

A is the area across which heat flows and Δt is the second surface temperature minus the first. For one homogeneous wall, such as X in figure 11.1, this becomes

$$Q = \frac{A(t_a - t_b)}{\dfrac{L_x}{\mathbf{k}_x}}$$

For a composite wall of several homogeneous substances, the form of (a) becomes

$$Q = \frac{A(-\Delta t)}{\sum L/\mathbf{k}} \text{ BTU/hr}$$

For example, $\sum L/\mathbf{k} = L_k/\mathbf{k}_x + L_y/\mathbf{k}_y$ for a two-section wall as in figure 11.1. Note that $L/\mathbf{k}$ is the thermal resistance per unit area of the wall L inches thick. Analogous to electrical resistances, these thermal resistances are added when they are in series. Conductivities $\mathbf{k}$ can be given in BTU-ft/ft^2-hr-°F or, more often, BTU-in/ft^2-hr-°F. L is either in inches or in feet to match $\mathbf{k}$. Check units in case of doubt. Also, the time unit is sometimes other than hours.

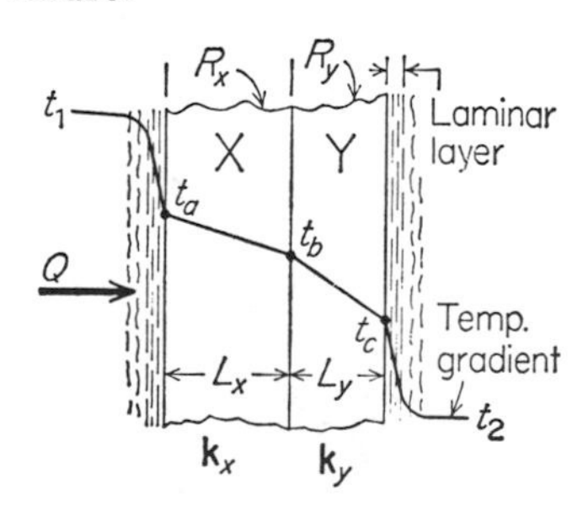

Figure 11.1

There is always a fluid film on the exposed surfaces (sometimes between sections) whose conducting properties are expressed by a *film coefficient* $\mathbf{h}$ BTU/hr-ft^2-°F that accounts for the expected film thickness. Resistance of a film is $1/\mathbf{h}$, and the over all resistance is $\sum R = \sum 1/\mathbf{h} + \sum L/\mathbf{k}$.

$$Q = \frac{A(t_1 - t_2)}{\sum R} = \frac{A(t_1 - t_2)}{\sum(1/\mathbf{h}) + \sum(L/\mathbf{k})} \text{ BTU/hr}$$

In heat conductance, the *transmittance*, U, also called the *over-all coefficient of heat transfer*, is used.

$$U = 1/\sum R \text{ BTU/ft}^2\text{-hr-°F}$$

(b)

$$\frac{1}{U} = \sum \frac{1}{\mathbf{h}} + \sum \frac{L}{\mathbf{k}} \ \frac{\text{ft}^2\text{-hr-°F}}{\text{BTU}}$$

Equation (a) then becomes

(c)

$$Q = UA(t_1 - t_2) = UA\Delta t_m \text{ BTU/hr}$$

Δt_m is defined by equation (e), below. Values of $\mathbf{k}$, $\mathbf{h}$, and U are found in texts and handbooks. A means of estimating $\mathbf{h}$ is given elsewhere in this chapter.

Curved Wall

In a pipe, for instance, the area across which heat flows changes with the distance from the center of the pipe. Integration of Fourier's equation gives the resistance of, say, the pipe or the insulation as

$$R = \frac{\ln(D_o/D_i)}{2\pi z\mathbf{k}}$$

D_o is the outside and D_i is the inside diameter of the section, z is the length (perpendicular to the direction of heat flow), and $\mathbf{k}$ is the conductivity of the material. Each fluid film has a different area whose resistance is $R = 1/(A\mathbf{h})$. For a pipe with a layer of insulation, the heat becomes

(d)

$$Q = \frac{t_1 - t_2}{\frac{1}{A_i\mathbf{h}_i} + \frac{\ln(D_b/D_a)}{2\pi z\mathbf{k}_x} + \frac{\ln(D_c/D_b)}{2\pi z\mathbf{k}_y} + \frac{1}{A_o\mathbf{h}_o}}$$

The subscript i is for inside, o for outside, and the diameters increase from D_a to D_c. D_b is the common diameter of the pipe and insulation. t_1 is the higher temperature, and t_2 is the lower temperature.

Logarithmic Mean Temperature Difference

In heat exchangers where cold and hot fluids are separated, the temperature difference $t_1 - t_2$ in equation (c) or (d) should be the logarithmic mean temperature difference Δt_m between the fluids.

$$\Delta t_m = \frac{\Delta t_A - \Delta t_B}{\ln(\Delta t_A/\Delta t_B)} = \frac{\Delta t_B - \Delta t_A}{\ln(\Delta t_B/\Delta t_A)} \qquad \text{(e)}$$

Δt_A is the difference in temperatures between the fluids at one end, and Δt_B is the temperature difference at the other end of the heat exchanger. Then, for example, $Q = UA(\Delta t_m)$. Equation (e) is good for counterflow or parallel flow exchanges and where one fluid remains at a constant temperature, as during condensation or evaporation.

RADIATION

Radiant heat Q_r is proportional to the fourth power of the absolute temperature. All bodies, therefore, radiate heat; the cooler body gains net heat. The Stefan-Boltzmann equation is

$$Q_r = 0.171 F_E F_A A \left[\left(\frac{T_1}{100} \right)^4 - \left(\frac{T_2}{100} \right)^4 \right] \text{ BTU/hr}$$

Q_r is radiant heat only, F_E is the *emissivity factor* and depends on the individual emissivities of the surfaces involved, F_A is the *angle factor* that allows for the average solid angle through which one surface "sees" the other, and A (square feet) is a characteristic area chosen to match the method of computing F_E, usually the smaller of two surfaces involved. Tables and charts are needed for choosing proper values of F_A, F_E, and A, or they must be given. For a small body 1 in a large enclosure,

$$F_E = \epsilon_1$$
$$F_A = 1$$
$$A = A_1$$

ϵ is the *emissivity.* Sometimes the radiant heat is small compared to the convected heat and is accordingly ignored (which is fairly safe when forced convection is used, but otherwise it might introduce serious inaccuracy).

CONVECTION

Film Coefficient

The *film coefficient* varies considerably with the circumstances, such as whether flow is laminar or turbulent, whether inside or outside of a pipe, whether or not one fluid (or both) is condensing (or evaporating), whether or not a double-pipe arrangement is used, and so forth. Most equations which give **h** are in terms of dimensionless groups (because most of them have been arrived at by dimensional analysis), such as the Reynolds number **R** and Prandtl number **P**. (See the chapter on fluid mechanics.) For turbulent flow inside of a pipe,

$$\frac{\mathbf{h}_i D}{\mathbf{k}_b} = 0.023 \left(\frac{D\mathcal{V}\rho}{\mu}\right)_b^{0.8} \left(\frac{c_p\mu}{\mathbf{k}}\right)_b^{0.4} \quad \text{(f)}$$

[NUSSELT EQUATION]

The subscript b means to use properties at the bulk temperature (say, the average of inlet and outlet temperatures of the fluid). $\mathbf{h}_i$ is the film coefficient for inner film, D is the internal diameter, **k** is the conductivity of the fluid, $\mathcal{V}$ is the average speed of fluid, ρ is the density, μ is the absolute viscosity, and c_p is the specific heat at constant pressure. Each of the groups in equation (f) is dimensionless:

$$\mathbf{N} = \frac{\mathbf{h}D}{\mathbf{k}}$$ [NUSSELT NUMBER]

$$\mathbf{R} = \frac{D\mathcal{V}\rho}{\mu}$$ [REYNOLDS NUMBER]

$$\mathbf{P} = \frac{c_p\mu}{\mathbf{k}}$$ [PRANDTL NUMBER]

Equation (f) is applicable when $0.7 < \mathbf{P} < 120$; for high-viscosity liquids, when $10{,}000 < \mathbf{R} < 120{,}000$; for gases and low-viscosity liquids, when $\mathbf{R} > 2100$; and when $L/D > 60$. In each number, the units must be consistent, but the system of units might be different for each number. Check units to be sure. In the Nusselt number **N**, the units cancel if **h** is in BTU/hr-ft^2-°F, D in inches, and **k** in BTU-in/hr-ft^2-°F. For the Prandtl number, recall the basic definitions of c_p (BTU/°F-unit mass, usually stated in tables as per pound of mass) and of μ. c_p contains the unit of mass and μ the unit of force. Thus, you must decide whether the pound is to be used for force or mass.

For ideal gases, the Prandtl number is given closely by an empirical equation:

(g)
$$\mathbf{P} = \frac{c_p \mu}{\mathbf{k}} \approx \frac{4}{9 - 5/k}$$

$k = c_p/c_v$. Since this number does not vary much for gases, an average value in equation (f) gives

$$\frac{\mathbf{h}_i D}{\mathbf{k}_b} = 0.021\mathbf{R}^{0.8}$$

This often is used for gases and low-pressure vapors.

EXAMPLE: A salt solution at the rate of 7000 lbm/hr is being heated from 60 °F to 190 °F by passing through a 150-foot copper tube, 1.2 inches in inside diameter. Outside the tube, water entering at 210 °F and leaving at 110 °F is circulated (countercurrent) at the rate of 7200 lbm/hr. The resistance of the copper tube to heat transfer is very small and can be neglected. For the salt solution, the Reynolds number is 39,000 and the Prandtl number is 6.15. Calculate (a) the over-all coefficient of heat transfer in BTU/hr-ft^2-°F based on the inside area of the tube, (b) the BTU transferred per foot of length of pipe per hour, (c) the length of exchanger required to double the capacity. The rates of flow of both liquids are to be doubled and thus the terminal temperatures are the same as before.

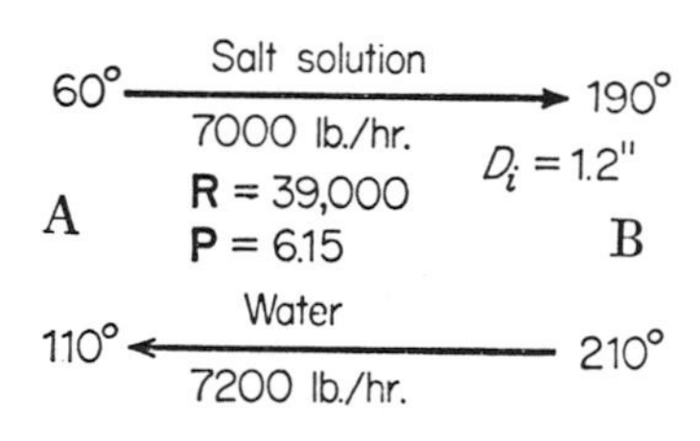

Figure 11.2

SOLUTION: (a) For convenient reference, make a sketch as in figure 11.2 with given data. To get the logarithmic temperature difference, use $\Delta t_A = 50$ and $\Delta t_B = 20$. From equation (e),

$$\Delta t_m = \frac{50 - 20}{\ln(50/20)} = 32.7°$$

Inside pipe area is

$$A = \pi DL = \pi(1.2/12)(150) = 47.1 \text{ ft}^2$$

The heat flow is computed from the water, using $c = 1$ BTU/lbm-°R:

$$Q = (7200)(210 - 110) = 720{,}000 \text{ BTU/hr}$$

Then, from equation (c), the value of U for the first part is

$$U_1 = \frac{Q}{A\Delta t_m} = \frac{720{,}000}{(47.1)(32.7)} = 468 \text{ BTU/hr-ft}^2\text{-°F} \quad (\textit{Ans.})$$

(b) $\dfrac{Q}{L} = \dfrac{720{,}000}{150} = 4800$ BTU/hr-ft length (*Ans.*)

(c) Since the details of the exchanger are not given, search for a solution from the given data. Observe that with the fluids moving faster, the heat transfer coefficient will be larger, so that doubling the capacity under this circumstance does not mean doubling the size of the exchanger.

$$Q_{\text{water}} = Q_{\text{salt}} = wc\Delta t = 7000c_s(130) = 720{,}000$$

From this, the specific heat for the salt solution is

$$c_s = 0.791 \text{ BTU/lbm-°F}$$

$$G = \frac{7000}{A_c} = \frac{7000}{\pi 1.2^2/(4 \times 144)} = 892{,}000 \text{ lbm/hr-ft}^2$$

This is the mass rate of flow per square foot. A_c is the cross-sectional area of tube. This value of G in Reynolds number gives μ for salt solution:

$$\mu = \frac{DG}{\mathbf{R}} = \frac{(1.2/12)(892{,}000)}{39{,}000} = 2.29 \text{ lbm/ft-hr}$$

Now the Prandtl number can be used to compute **k**.

$$\mathbf{k} = \frac{c_p\mu}{\mathbf{P}} = \frac{(0.791)(2.29)}{6.15} = 0.295 \text{ BTU-ft/ft}^2\text{-hr-°F}$$

(Check the balance of all units.) The Nusselt equation (f) can next be used to determine $\mathbf{h}_i$ for the salt solution. (Somewhat different equations might be found, depending on the reference used.)

$$\frac{\mathbf{h}_i(1.2/12)}{0.295} = 0.023(39{,}000)^{0.8}(6.15)^{0.4} = 224$$

From this,

$$\mathbf{h}_{i1} = 661 \text{ BTU/ft}^2\text{-hr-°F}$$

From the computed U_1 in part (a) and equation (b) above, letting L/k be zero because it is small, get $\mathbf{h}_{o1}$.

$$\frac{1}{U_1} = \frac{1}{\mathbf{h}_{i1}} + \frac{1}{\mathbf{h}_{o1}} = \frac{1}{468} = \frac{1}{661} + \frac{1}{\mathbf{h}_{o1}}$$

From this,

$$\mathbf{h}_{o1} = 1062 \text{ BTU/ft}^2\text{-hr-°F}$$

This is the film coefficient on the outside of the tube.

Consider each term in the Nusselt equation, and note that all properties remain the same except that the velocity must double in order to get the same amount of fluid through the same size area. Thus, $\mathbf{h}_i$ varies as $\mathcal{V}^{0.8}$ (given the exponents quoted above).

$$\mathbf{h}_{i2} = \left(\frac{\mathcal{V}_2}{\mathcal{V}_1}\right)^{0.8} \mathbf{h}_{i1} = 2^{0.8}(661) = 1150 \text{ BTU/ft}^2\text{-hr-°F}$$

Determination of $\mathbf{h}_{o2}$ is subject to considerable question. The only reasonable thing to do is to assume that the space outside of the tube has an annulus section, in which case $\mathbf{h}_o$ is a function of $\mathbf{R}^{0.8}$, same as for the inside of the tube. (Some other reference might yield another exponent.) Therefore, as for $\mathbf{h}_i$,

$$\mathbf{h}_{o2} = \left(\frac{\mathcal{V}_2}{\mathcal{V}_1}\right)^{0.8} \mathbf{h}_{o1} = 2^{0.8}(1062) = 2790 \text{ BTU/ft}^2\text{-hr-°F}$$

$$\frac{1}{U_2} = \frac{1}{\mathbf{h}_{i2}} + \frac{1}{\mathbf{h}_{o2}} = \frac{1}{1150} + \frac{1}{2790}$$

$$U_2 = 815 \text{ BTU/hr-ft}^2\text{-°F}$$

Area needed is

$$A = \frac{Q}{U\Delta t_m} = \frac{1.44 \text{ EE6}}{(815)(32.7)} = 54 \text{ ft}^2 = \pi D L$$

$$L = 173 \text{ ft} \quad (\mathit{Ans.})$$

PRACTICE PROBLEMS

11.1. The temperature under a floor is 35 °F, and the room temperature is 70 °F. Under these conditions, 15 BTU/hr-ft^2 are lost through the floor in figure 11.3(a). In figure 11.3(b), insulation is added to the floor, and the conductances for the materials are: film coefficient for room to floor = 1.21 BTU/hr-ft^2-°F; conductance through floor board = 1.02 BTU/hr-ft^2-°F; conductance through insulation = 0.15 BTU/hr-ft^2-°F; film coefficient for insulation surface to under-floor air = 1.21 BTU/hr-ft^2-°F. (a) What is the magnitude of the over-all heat transfer coefficient for (a)? (b) What is the magnitude of the over-all heat transfer coefficient for (b)? (c) What is the magnitude of the heat lost through the insulated floor in BTU/hr-ft^2?

Ans. (a) 0.38 BTU/hr-ft^2-°F, (b) 0.108 BTU/hr-ft^2-°F, (c) 3.78 BTU/hr-ft^2

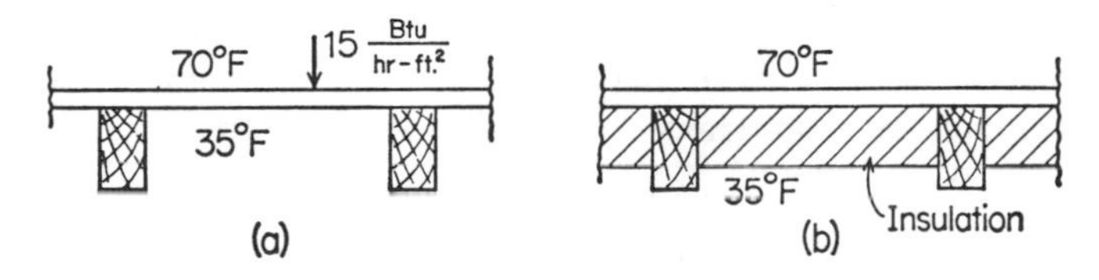

Figure 11.3

11.2. The following coefficients are given: film coefficient for indoor surface = 1.65, plaster = 4.64, brick = 0.63; film coefficient for outdoor surface = 6.0, each in BTU/hr-ft^2-°F. (a) Calculate the over-all heat transfer coefficient for the wall as shown in figure 11.4. (b) If the indoor temperature is 70 °F, and the outdoor temperature is 28 °F, how much heat (in BTU/hr) will pass through 150 square feet of wall area?

Ans. (a) 0.388 BTU/hr-ft^2-°F, (b) 2440 BTU/hr

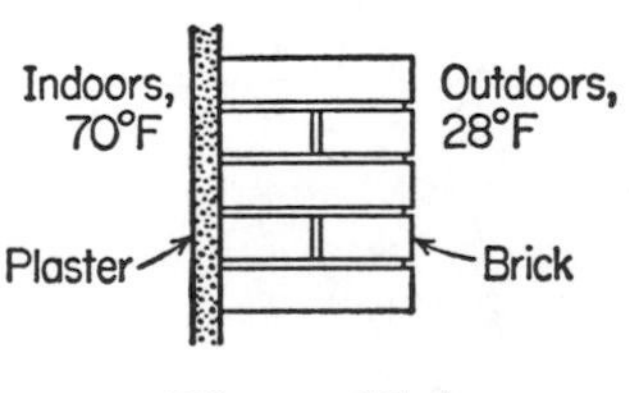

Figure 11.4

11.3. If all other factors remain the same, heat radiated from a unit surface: (a) varies directly as the surface temperature in °F, (b) varies inversely as the surface temperature in °F, (c) varies directly as the fourth power of the absolute temperature of the surface, (d) varies inversely as the fourth power of the absolute temperature of the surface, (e) varies directly as the square of the absolute temperature of the surface.

Ans. (c)

11.4. (a) Find the rate of heat flow through a composite wall composed of 4 inches of brick (**k** = 0.40 BTU/hr-ft-°F), 12 inches of concrete (**k** = 0.70 BTU/hr-ft-°F), and 2 inches of plaster (**k** = 0.55 BTU/hr-ft-°F). The film coefficient for heat transfer for the outside surface of the wall ($\mathbf{h}_o$) is 4.0 BTU/hr-ft^2-°F, and for the inside surface of the wall ($\mathbf{h}_i$) is 1.2 BTU/hr-ft^2-°F. The outside air temperature is 35 °F, and the inside air temperature is 70 °F. Give the answer per square foot of wall surface. (b) Find the temperature of the outside surface of the wall.

Ans. (a) 9.6 BTU/hr-ft^2, (b) 37.4 °F

11.5. A single-pass shell and tube exchanger operating counter-currently has a temperature difference on the hot end of 20 °F and on the cold end of 10 °F. Calculate: (a) the long mean temperature difference, (b) and the rate of heat transfer in BTU/hr-ft^2 if the over-all transfer coefficient is 50 BTU/hr-ft^2-°F.

Ans. (a) 14.4 °F, (b) 720 BTU/hr-ft^2

12 *ELECTRICITY*

SYMBOLS

A = area
B = magnetic flux density ($= \phi/A$)
C = capacitance, farad (F), microfarad (μF)
D = diameter
E = potential difference, volts
e = instantaneous potential difference, AC
f = frequency ($1/T$), hertz (Hz)
G = conductance, mho, reciprocal of resistance
I = current, amperes (amp), Q/t
i = instantaneous current, AC
J = operator
L = inductance, henrys
l = length
M = mutual inductance, henrys
MMF = magnetomotive force
n = speed (rpm)
N = number of turns of wire in a coil
P = power, watt (W), kW, etc.
Q = quantity of electricity, coulomb
R = resistance, ohm (Ω)
$\mathcal{R}$ = reluctance
T = period ($1/f$), seconds/cycle
t = time
V = line or terminal voltage
v = linear speed
X_L = inductive reactance, ohm
X_C = capacitive reactance, ohm

Y = admittance, mho
Z = impedance, ohm
γ = electrical conductivity, mho/cm
μ = permeability
ρ = electrical resistivity (often ohm-cm), electrical energy (erg, joule)
ω = angular velocity ($2\pi f$)
Ω = symbolic abbreviation of ohm
ϕ = magnetic flux

As in other chapters, emphasis is placed on ideas and principles that a graduate of any branch of engineering might be expected to know.

DIRECT CURRENT

Ohm's Law

The potential difference (E volts) between two points in an electrical circuit is equal to the product of the current flowing (I amp) and the resistance (R ohms) between the points:

(a)
$$E = IR$$
$$I = \frac{E}{R}$$
$$R = \frac{E}{I}$$

Electrical energy, ρ, is 100 percent available energy. Electrical energy can be completely converted to shaft work, and vice versa, without loss in an ideal machine. Power (P watts), the rate of doing work, is $P = EI$. Using equation (a),

(b)
$$P = EI = I^2R = E^2/R \text{ watts}$$

A watt is a unit of power. A watt-hour is a unit of energy. A kilowatt-hour is 1000 watt-hr; 746 watts is equivalent to one horsepower.

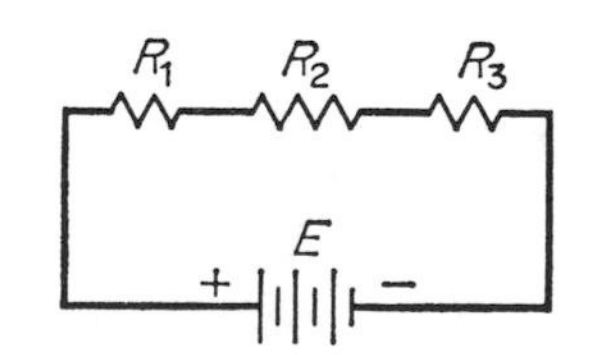

Figure 12.1

If resistances are connected in *series*, as in figure 12.1, The current I through each part of the circuit is the same and the total resistance is the sum of the R's of the parts:

$$R = R_1 + R_2 + R_3 \text{ ohms} \tag{c}$$

The voltage drop across each resistance is IR, which is the sum of those of the parts:

$$E = IR = IR_1 + IR_2 + IR_3 \text{ volts} \qquad \text{[SERIES]} \tag{d}$$

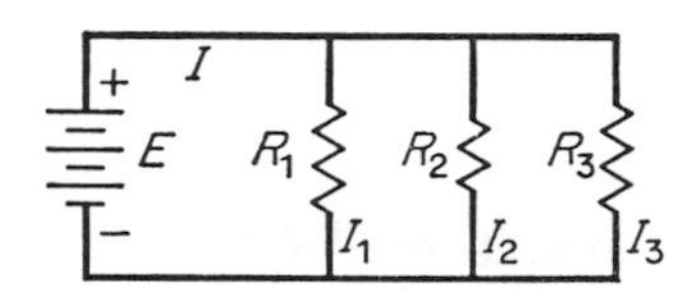

Figure 12.2

If resistances are connected in *parallel*, as in figure 12.2, the voltage is the same across each branch. Hence, the current in each branch is inversely proportional to the resistance: $I_1 = E/R_1$, $I_2 = E/R_2$, and so forth. The total current I is the sum:

$$I = \frac{E}{R} = I_1 + I_2 + I_3 = \frac{E}{R_1} + \frac{E}{R_2} + \frac{E}{R_3} \tag{e}$$

$$I = E\left(\frac{1}{R_1} + \frac{1}{R_2} + \frac{1}{R_3}\right) \text{ amps} \tag{f}$$

$$\frac{1}{R} = \frac{1}{R_1} + \frac{1}{R_2} + \frac{1}{R_3} \text{ ohms} \qquad \text{[PARALLEL]}$$

R is the net or equivalent resistance of the circuit; R is less than R_1, R_2, or R_3. Also,

$$I_1R_1 = I_2R_2 = I_3R_3 = E \text{ volts} \tag{h}$$

Resistivity ρ, a property of a material, is the DC resistance of a portion that has unit area and unit length:

$$\rho = RA/l$$
$$R = \rho l/A$$

l is the length of the conductor whose cross-sectional area is A.

Quantity of electricity is measured in coulombs, Q. 1 amp = 1 coulomb/sec, and $I = Q/t$.

Circular mil (CM), used for wire measure, is the area of a circle whose $D = 0.001$ inch; CM $= \pi 0.001^2/4$ in^2, or, for conversion purposes,

$$\frac{\pi}{4\ \text{EE6}} \frac{\text{in}^2}{\text{CM}}$$

$$\frac{4\ \text{EE6}}{\pi} \frac{\text{CM}}{\text{in}^2}$$

Resistivity is often given in units of ohm-CM/ft.

Kirchhoff's Laws

1. If a number of wires connect at a particular point, the DC current flowing to the point is equal to that flowing from the junction.

2. The total change of voltage around any closed circuit is zero. Currents flowing toward the junction are positive, and those flowing from the junction are negative. Rise in voltage is positive (through *EMF* source from negative to positive, through a resistance *in opposite sense* to current). Drop in voltage is negative (through *EMF* source from positive to negative, through a resistance in the same sense as the current).

EXAMPLE 1: A ribbon of copper having a resistivity of 10.4 ohm-CM/ft is 4 meters long, 0.06 inch wide, and 0.02 inch thick. A 6-volt battery is connected to its ends through two wire leads, each having a resistance of 0.15 ohms. Compute the voltage developed between the ends of the ribbon.

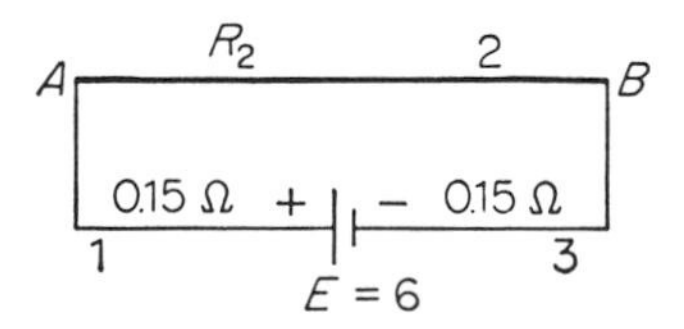

Figure 12.3

SOLUTION: The area of the ribbon is

$$A = (0.06)(0.02)\left(\frac{4\ \text{EE6}}{\pi}\right) = 1530\ \text{CM}$$

For $l = 4$ meters $= (4)(3.28) = 13.12$ feet,

$$R_2 = \frac{\rho l}{A} = \frac{(10.4)(13.12)}{1530} = 0.0893\ \text{ohm}$$

The circuit is represented in figure 12.3.

$$\text{total } R = R_1 + R_2 + R_3 = 0.30 + 0.0893 = 0.3893\ \text{ohms}$$
$$I = E/R = 6/0.3893 = 15.41\ \text{amp}$$

Then, the voltage across AB is

$$E_{AB} = IR_2 = (15.41)(0.0893) = 1.38\ \text{volts} \quad (\textit{Ans.})$$

EXAMPLE 2: The voltage and internal resistance of two batteries are 20 volts, 0.25 ohm, and 19 volts, 0.40 ohm, respectively. The batteries are paralleled to supply a common load having a resistance of 0.5 ohm. Calculate the current and power taken by the load.

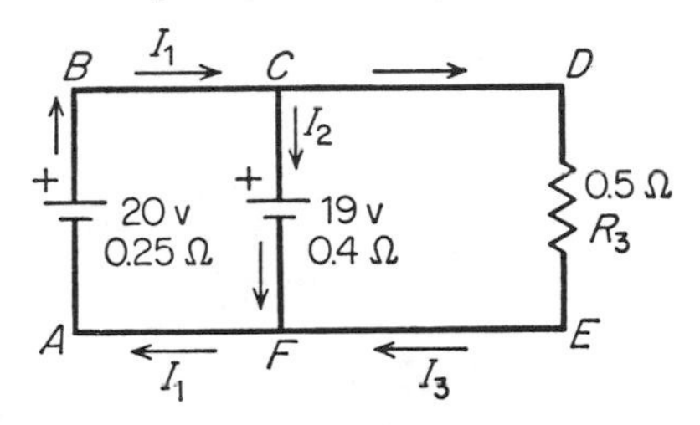

Figure 12.4

SOLUTION: Sketch the circuit, as in figure 12.4, and assume directions of current flow. If wrong directions are chosen, a negative sign in the final result will so indicate. Apply Kirchhoff's laws. At point C (or F),

(i) $$I_1 - I_2 - I_3 = 0$$

For loop $ABDE$,

(j) $$\Delta E = 20 - 0.25I_1 - 0.5I_3 = 0$$

For loop *CDEF*,

(k) $$\Delta E = -0.5I_3 + 19 + 0.4I_2 = 0$$

Solve (i), (j), and (k) simultaneously and find $I_3 = +30$ amp (shown in correct direction). (*Ans.*)

Check your work as far as possible. This can be done by computing I_1 and I_2. (I_2 is found to be in the wrong direction.) Apply $\Delta E = 0$ to the *ABCF* loop.

The electrical power being dissipated as heat via the load R_3 is

$$P = I^2R = (30)^2(0.5) = 450 \text{ W} \quad (\textit{Ans.})$$

MAGNETISM

Magnetomotive force MMF (not a force, but analogous to *EMF* in an electric circuit) produces or tends to produce magnetic flux, and is measured in *gilberts* (cgs system) and *ampere-turns NI* (rationalized mks system):

$$MMF = 0.4\pi NI \text{ gilberts}$$

N is the number of turns of coil and I is the amperes of current.

Magnetic flux ϕ (analogous to current in an electric circuit) is the magnetic "flow" in a magnetic circuit. It is measured in webers (mks) or in *maxwells* or *lines* (cgs). There are 1 EE8 maxwells per weber.

In a magnetic circuit, *reluctance* $\mathcal{R}$ is analogous to electric resistance R:

(l) $$MMF = \phi\mathcal{R}$$

$$\phi = \frac{MMF}{\mathcal{R}}$$

$$\mathcal{R} = \frac{MMF}{\phi}$$

If ϕ is in lines or maxwells and *MMF* is in ampere-turns, $\mathcal{R}$ is

$$\mathcal{R} = \frac{l}{A\mu}$$

l cm is the length of the part of the circuit whose cross-sectional area is A square centimeters, and μ is the *permeability.* Handle reluctances in parallel and series in the same manner as resistances. *Magnetic flux density* B is the magnetic flux per unit area. 1 gauss $=$ 1 maxwell/cm^2 $=$ 1 line/cm^2, and $B = \phi/A$. For direction of "flow," use the "right-hand rule": Point the outstretched thumb in the direction of the current I and the direction in which the fingers point around the electrical conductor is the direction of the magnetic flux (figure 12.5).

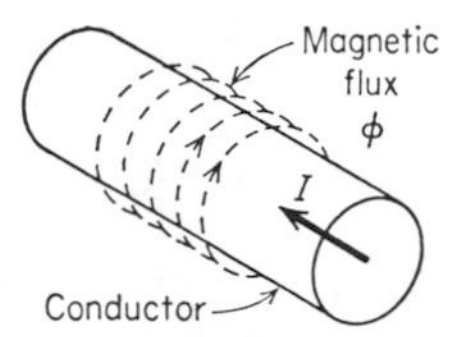

Figure 12.5

DIRECT CURRENT MACHINERY

If an electrical conductor of length l inches moves across a magnetic field of strength B lines/in^2 at the rate of v inches/second, the induced *EMF* in a motor or generator is

$$E = Blv \times 10^8 \text{ volts}$$

For a particular machine, this is seen to vary directly as the speed and flux.

$$E = K\phi n \text{ volts}$$

K is a constant and n = rpm. If E is the induced voltage, the terminal voltage of a shunt dynamo is

$$V = E \pm I_a R_a$$

(Sign is + for a motor, − for a generator.) I_a is the armature current and R_a is the total resistance of the armature between terminals. The field circuit might be in parallel with the armature circuit, the shunt winding as in figure 12.6(a), in series, as in figure 12.6(b), or a combination, as in figure 12.6(c).

The induced voltage is practically directly proportional to field current—up to the point where the iron of the magnetic circuit begins to become saturated.

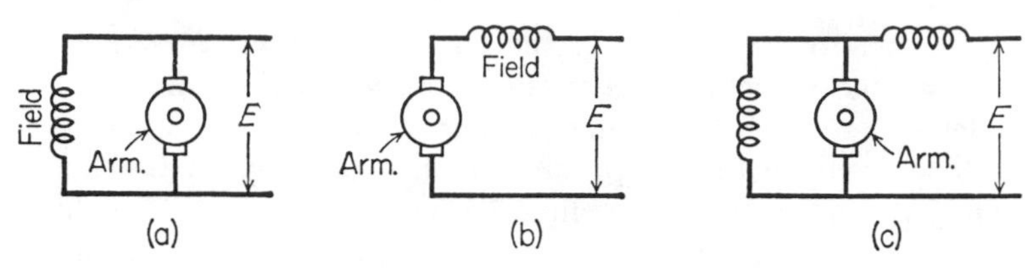

Figure 12.6

Motor Torque

Motor torque is proportional to flux density ϕ and armature current I_a:

$$T = K'\phi I_a$$

K', a constant for a particular motor, is a function of the number and length of active conductors. Also, $T = 33{,}000 \times \text{hp}/\omega$ ft-lbf (from mechanics) for ω radians/minute angular velocity.

Motor Speed

Where speed control is desired, a series motor has

$$n = \frac{V - I_a(R_a + R_s)}{K'' I_a} = \frac{E_m}{K\phi} \text{ rpm}$$

V is terminal (line) voltage, K'' is a constant for a particular machine. I_a and R_a refer to armature, and R_s refers to series field. E_m is induced voltage or counter-electromotive force.

$$E_m = V - I_a(R_a + R_s)$$
$$K\phi = K'' I_a$$

If saturation is neglected, the flux is proportional to the current.

The *efficiency* of a motor is the power output divided by the power input.

ALTERNATING CURRENTS

Current and voltage vary almost sinusoidally, as in figure 12.7. Across a resistor, current and voltage are in phase, both zero and maximum at the same instant. Instantaneous values are:

$$i = I_m \sin\theta = I_m \sin 2\pi f t$$

(n) $$e = Ri = RI_m \sin 2\pi f t = E_m \sin 2\pi f t$$

f (Hz) is the frequency (analogous to n rps) and $2\pi f$ radians/second $= \omega$. I_m and E_m are maximum values. Instantaneous power is

$$p = Ri^2 = RI_m^2 \sin^2 2\pi ft$$

Average power is

$$P = \frac{RI_m^2}{2} \quad \text{(o)}$$

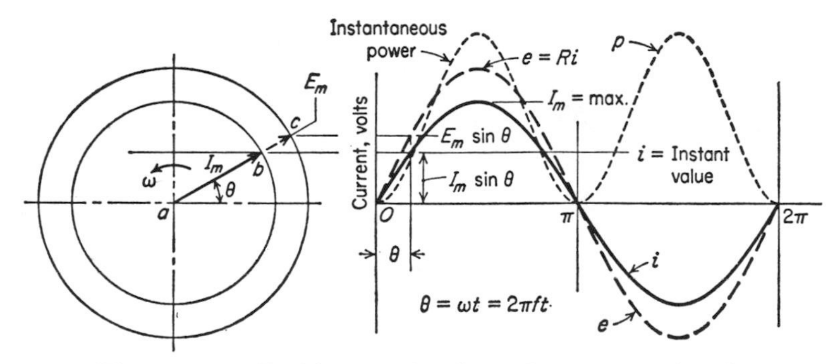

Figure 12.7 Positive rotation for radius vectors *ab* and *ac* is counterclockwise at ω radians/sec.

The effective alternating current, which is the root mean square (rms), is an amount that would produce the same heat in a resistor as a direct current of the same number of amperes.

$$I_{\text{eff}} = I_{\text{DC}} = I$$

Average power is

$$P = I_{\text{eff}}^2 R = I^2 R = \frac{I_m^2 R}{2}$$

$$I = \frac{I_m}{\sqrt{2}}$$

$$E_{\text{eff}} = E = \frac{I_m R}{\sqrt{2}} = IR$$

This is true for sinusoidal voltage and current. The rate of change of current is

$$\frac{di}{dt} = I_m 2\pi f \cos 2\pi ft$$

Inductive Circuits

The *inductance*, L henrys, is related to voltage in

$$e = -L\,di/dt = -LI_m 2\pi f \cos 2\pi f t$$

In this, the *inductive reactance*, X_L, of a coil is

$$X_L = 2\pi f L = \omega L \text{ ohms}$$

Related to effective current I and voltage E,

(p)
$$E = X_L I$$
$$E_{\text{max}} = X_L I_m \text{ volts}$$

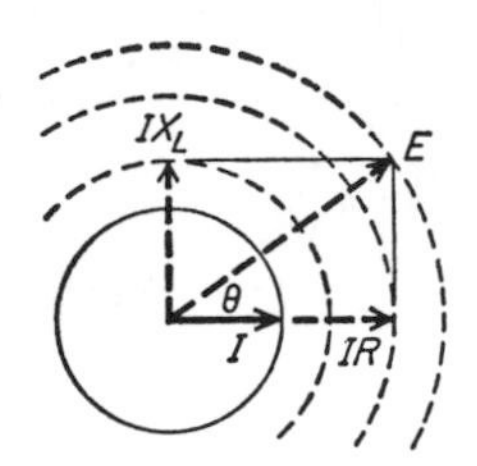

Figure 12.8

This current lags the voltage 90°. The situation in a coil with a resistance R and an inductive reactance X_L in series is shown in figure 12.8. E is the vector sum of IR and IX_L:

(q)
$$\mathbf{E} = \mathbf{I}(R + jX_L) = \mathbf{IZ}$$

The boldface symbols signify vector properties. j is the operator that says that $\mathbf{X}_L$ is at right angles to $\mathbf{R}$, and $\mathbf{Z} = R + jX_L$ is the *impedance.* The magnitude of $\mathbf{Z}$ is

$$\mathbf{Z} = \sqrt{R^2 + X_L^2} \quad \text{at } \theta = \tan^{-1}\frac{X_L}{R}$$

θ is the angle that the voltage leads the current. *Admittance*, Y, is the reciprocal of impedence, Z:

$$Y = \frac{1}{Z}$$
$$\mathbf{I} = \mathbf{EY}$$

Capacitive Circuits

Capacitive reactance, X_c, is

$$X_c = \frac{1}{2\pi fC} = \frac{1}{\omega C} \text{ ohms}$$

C farads is the capacitance, usually given in microfarads, μF, or micromicrofarads, $\mu\mu$F. There are 1 EE6 μF per F, and 1 EE12 $\mu\mu$F per F. For capacitors in parallel,

$$C = C_1 + C_2 + C_3 + \ldots$$

For series connected capacitors,

$$\frac{1}{C} = \frac{1}{C_1} + \frac{1}{C_2} + \frac{1}{C_3} + \ldots$$

The voltage across a capacitive reactance is

$$E = IX_c = \frac{I}{2\pi fC}$$

$$I = \frac{E}{X_c} = 2\pi fCE$$

I and E are effective (rms) values and *the current leads the voltage* by 90°, as in figure 12.9. The capacitive admittance is

$$Y = 1/Z = 2\pi fC$$

For a capacitive reactance in series with a resistor, the impedance is

$$Z = \sqrt{R^2 + X_C^2} = \sqrt{R^2 + \left[\frac{1}{2\pi fC}\right]^2} \text{ ohms}$$

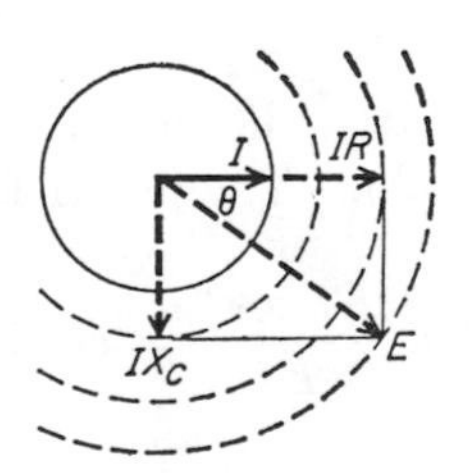

Figure 12.9

In vector form,

$$\mathbf{E} = \mathbf{I}(R - jX_C) = \mathbf{IZ}$$

The acute negative angle by which voltage lags current is

$$\theta = \tan^{-1}(-IX_C/R)$$

To get admittance **Y** from **Z** in vector form,

$$\mathbf{Y} = \frac{1}{\mathbf{Z}} = \frac{1}{Z\underline{/-\theta}} = \frac{1}{Z}\underline{/\theta} = Y\underline{/\theta} \text{ mhos}$$

This is useful in problems for parallel circuits. The same voltage drop occurs across parallel parts. For resistance, inductance, and capacitance in series,

$$Z = \sqrt{R^2 + (X_L - X_C)^2} \text{ ohms at } \theta = \tan^{-1}\frac{X_L - X_C}{R}$$

The same current flows through each, $\mathbf{I} = \mathbf{E}/\mathbf{Z}$, and it leads or lags the voltage according to whether X_C is greater than or less than X_L, as in figure 12.10.

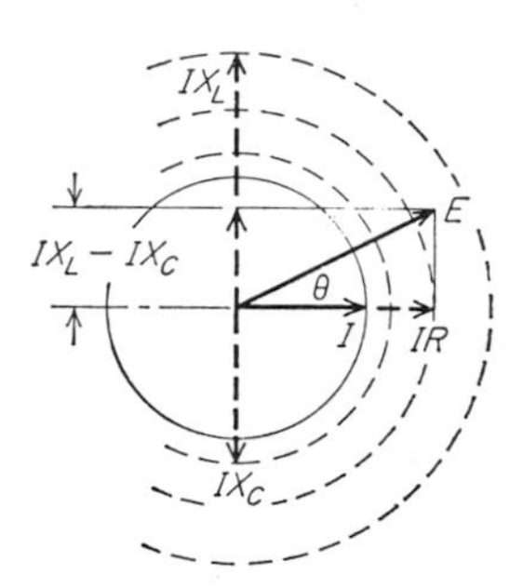

Figure 12.10

Power Factor

True power equals $EI\cos\theta$ watts, the in-phase voltage and current. Apparent power equals EI volt-amp. The power factor is

$$\text{pf} = \frac{EI\cos\theta}{EI} = \cos\theta$$

$E\cos\theta = IR$, as in figure 12.10.

Resonance

Resonance occurs when $X_L = X_C$ or $2\pi f L = \frac{1}{2\pi f C}$, from which the resonant frequency is

$$f = 1/(2\pi\sqrt{LC})$$

The current and power are a maximum in a series circuit.

$$Z = R$$

(The current is a minimum in a parallel circuit.)

EXAMPLE: Two impedance coils are connected in parallel. Coil A has a resistance of 3 ohms and an inductive reactance of 4 ohms. Coil B takes 12 amp and 2500 watts from a 220-volt, 60 Hz AC line. (a) How much capacitance must be connected in series with the parallel connected coils so that the circuit will have unity power factor when operating with a line voltage of 220 volts at 60 Hz? (b) With unity power factor, how much power is taken from the source given in (a)?

SOLUTION: (a) First, sketch the circuit, as in figure 12.11. Then find the impedance of coil B at 220 volts, 60 Hz condition. The in-phase resistance is

$$R_B = \frac{P}{I^2} = \frac{2500}{12^2} = 17.36 \text{ ohms}$$

The impedance, first condition, is

$$Z_{B1} = \frac{E}{I} = \frac{220}{12} = 18.33 = R + jX_{B1} = 17.36 + jX_{B1} \text{ ohms}$$

From this,

$$\begin{aligned} X_{B1} &= 5.92 = 2\pi f L_B = 2\pi(40)L_B \\ L_B &= 0.0236 \text{ henrys} \\ X_{B2} &= 2\pi(60)(0.0236) = 8.88 \text{ ohms} \\ Z_{B2} &= 17.36 + j8.88 \text{ ohms} \end{aligned}$$

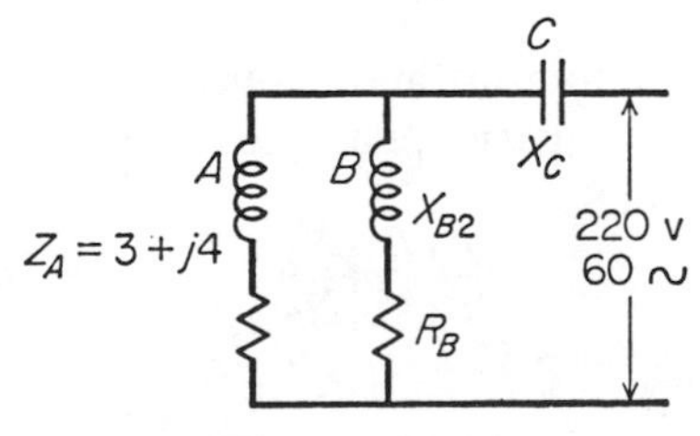

Figure 12.11

For parallel circuits A and B in figure 12.11, transforming to admittances,

$$\mathbf{Y}_A = \frac{1}{\mathbf{Z}_A} = \frac{1}{3+j4} = \frac{1}{5/\underline{+36.9^\circ}} = 0.2/\underline{-36.9^\circ}$$

$$\mathbf{Y}_A = 0.16 - j0.12 \text{ mhos}$$

$$\mathbf{Y}_B = \frac{1}{\mathbf{Z}_{B2}} = \frac{1}{17.36+j8.88} = \frac{1}{19.49/\underline{+27.2^\circ}} = 0.0513/\underline{-27.2^\circ}$$

$$\mathbf{Y}_B = 0.0456 - j0.0237 \text{ mhos}$$

$$\mathbf{Y}_{AB} = Y_A + Y_B = 0.2056 - j0.1437 = 0.251/\underline{-35^\circ}$$

$$\mathbf{Z}_{AB} = \frac{1}{\mathbf{Y}_{AB}} = \frac{1}{0.251/\underline{-35^\circ}} = 3.99/\underline{+35^\circ}$$

$$= 3.27 + j2.28 \text{ ohms}$$

$$Z_C = -jX_C$$

$$Z_{ABC} = 3.27 + j2.28 - jX_C$$

For power factor of unity,

$$j2.28 = jX_C$$

$$X_C = 2.28 \text{ ohms}$$

$$C = \frac{1}{2\pi f X_C} = \frac{1}{2\pi(60)(2.28)} = 1160\ \mu\text{F} \quad \text{(Ans.)}$$

(b) With pf = 1, the power flowing through the system is

$$P = I^2R = E^2/R = 220^2/2.28 = 21{,}250 \text{ W or } 21.15 \text{ kW}$$

Transformers

With sinusoidal wave form, the *EMF* in one coil is

$$E = 4.44\phi_m N f(1 \text{ EE–8}) \text{ volts}$$

ϕ_m maxwells is the maximum instantaneous flux. N is turns on the coil. f is frequency. From this equation, for sides 1 and 2 of a transformer,

$$\frac{E_1}{E_2} = \frac{N_1}{N_2}$$

$$N_1 I_1 = N_2 I_2$$

Vacuum Tube

The triode tube in figure 12.12 consists of a plate with the positive connection, a filament, and a grid. With *EMF* applied, electrons (negative) move from filament to plate (i.e., electricity "flows" from plate to filament). Small changes of potential between grid and filament or cathode control the rate of flow of electrons.

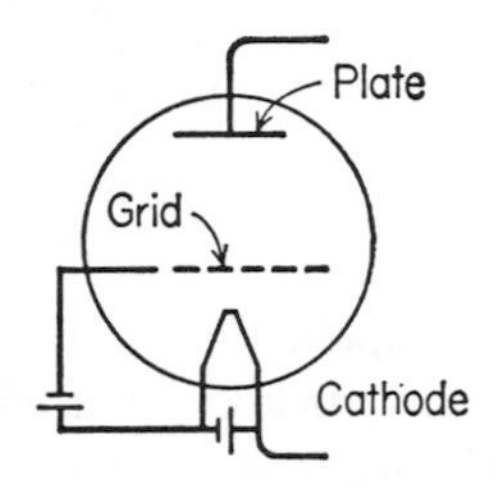

Figure 12.12

Conversion Constants

$1\ \frac{\text{amp-sec}}{\text{coulomb}}$	$1\ \frac{\text{newton-meter}}{\text{joule}}$	$1\ \frac{\text{joule}}{\text{watt-sec}}$
$100{,}000\ \frac{\text{dyne}}{\text{newton}}$	$1054.8\ \frac{\text{joule}}{\text{BTU}}$	$1\text{ EE–7}\ \frac{\text{joule}}{\text{erg}}$
$1000\ \frac{\text{watt}}{\text{kW}}$	$0.1\ \frac{\text{abamp}}{\text{amp}}$	$360\ \frac{\text{abcoulomb}}{\text{amp-hr}}$
$1.257\ \frac{\text{gilbert}}{\text{amp-turn}}$	$1\text{ EE6}\ \frac{\text{ohm}}{\text{megohm}}$	$0.001\ \frac{\text{volt}}{\text{millivolt}}$
$1\text{ EE6}\ \frac{\mu\text{F}}{\text{F}}$	$1\text{ EE–8}\ \frac{\text{weber}}{\text{cm}^2\text{-gauss}}$	$1\text{ EE8}\ \frac{\text{maxwell}}{\text{weber}}$
$0.001\ \frac{\text{henry}}{\text{millihenry}}$	$1\text{ EE6}\ \frac{\text{microhenry}}{\text{henry}}$	$1.273\text{ EE6}\ \frac{\text{circular mil}}{\text{in}^2}$

PRACTICE PROBLEMS

12.1. If the current from a short-circuited 1.5-volt dry cell is 25 amp, what is the internal resistance of the cell? The resistance of the short circuit is negligible. If this cell is connected through a 1-ohm external circuit, find the current and the reading of a voltmeter connected to its terminals.

Ans. 0.06 ohm, 1.414 amps, 1.415 v

12.2. An electric truck is propelled by a DC motor drive connected to a 110-volt storage battery. This truck is required to exert a tractive effort of 200 lbf at a speed of 5 mph. If the over-all efficiency of motor and drive is 70%, what current is taken from the battery?

Ans. 25.8 amps

12.3. For the circuit shown in figure 12.13, what current will flow through the resistance R_1?

Ans. 0.5 amps

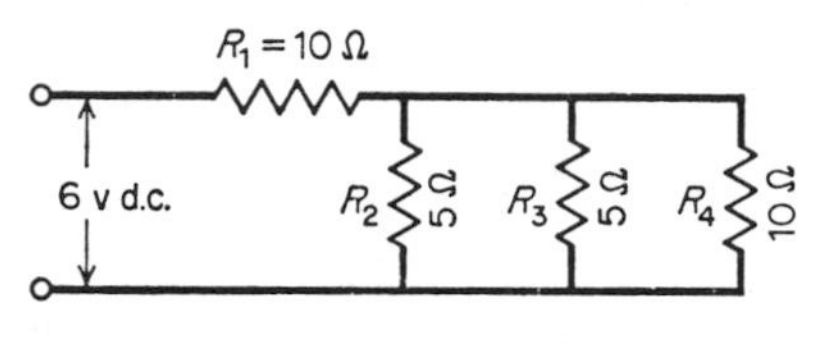

Figure 12.13

12.4. Show how to connect a double-throw, double-pole switch to a direct current shunt-wound motor with a separately-exited field so that the motor can be reversed by the switch. See figure 12.14.

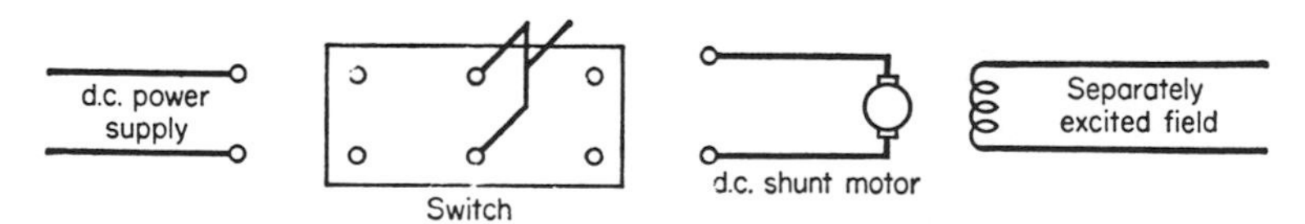

Figure 12.14

12.5. The rated horsepower of an electric motor is: (a) the horsepower of the electric energy necessary to operate at full load, (b) the horsepower delivered to the drive shaft at twice the rated current, (c) the horsepower necessary to twist the shaft out of shape, (d) the horsepower output that causes a specified rise in temperature when the motor is run continuously.

Ans. (d)

12.6. The magnitude of a voltage generated by a single conductor passing through a magnetic field is *not* a function of: (a) the speed the conductor has in cutting the magnetic field, (b) the magnetic field density, (c) the diameter of the conductor cutting the magnetic field, (d) the length of the conductor cutting the magnetic field, (e) any of these.

Ans. (c)

12.7. A 6-volt battery is connected to the resistances shown in figure 12.15. Neglecting all resistance except as shown, what current will flow through each resistance?

Ans. 1.5 amps, 2.0 amps

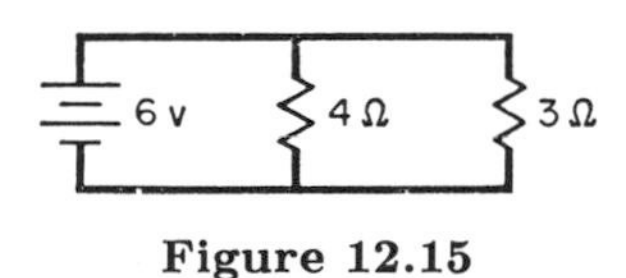

Figure 12.15

12.8. Which costs more to operate, a 5-hp (input) motor used 1 hour every other day for 30 days or a 500-watt lamp used 4 hours each day for 30 days?

Ans. lamp

12.9. A DC 0 to 50 millivoltmeter is used to measure 5 amp line current by means of a shunt. The meter was tested and found to have a terminal resistance of 2.17 ohms. (a) How many ohms resistance must be connected in parallel with the meter so that it will indicate a full-scale deflection at 5.00 amp line current? (b) How many amperes will pass through the shunt resistor at full-scale deflection? (c) By what factor must the reading on the meter be multiplied in order to get the true line current?

Ans. (a) 0.0101 ohms, (b) 4.977 amps, (c) 100

12.10. (a) A current of 1 amp at 50 volts is required to feed an electric lamp. What is the resistance of the lamp in ohms? (b) How many hours must the lamp burn in order to consume one kilowatt-hour? Assume unity power factor. (c) What will be the energy consumption if two of these lamps are connected in series across 50 volts? Assume that the resistance of the lamps does not change with temperature.

Ans. (a) 50 ohms, (b) 20 hrs, (c) 25 watts

12.11. A DC generator has an armature resistance of 0.31 ohm. The shunt field has a resistance of 134 ohms. No-load voltage is 121 volts at 1775 rpm. Full-load current is 30 amp at 110 volts. Find: (a) field current at no load, (b) speed at full load, (c) stray power losses if efficiency at full load is 78%.

Ans. (a) 0.903 amp, (b) 1930 rpm (assuming that flux varies directly with field current), (c) 545 watts.

12.12. 20 cells of 1.5 volts each and internal resistance of 0.1 ohm each will deliver how much current through a circuit with resistance of 0.5 ohm (a) if cells are connected in series? (b) if cells are connected in parallel? (c) if a single cell is connected in a circuit with 0.5 ohm resistance?

Ans. (a) 12 amps, (b) 2.97 amps, (c) 2.5 amps

12.13. A four-pole generator is turned at 3600 rpm. Each of its poles has a flux of 30,000 lines. The armature has 200 conductors and is simplex lap wound. Approximately what voltage is generated at no load?

Ans. 3.6 volts

12.14. Find the voltage and current of each resistor in the circuit shown in figure 12.16.

Ans. 1 amp through R_3, 20 volts across R_6

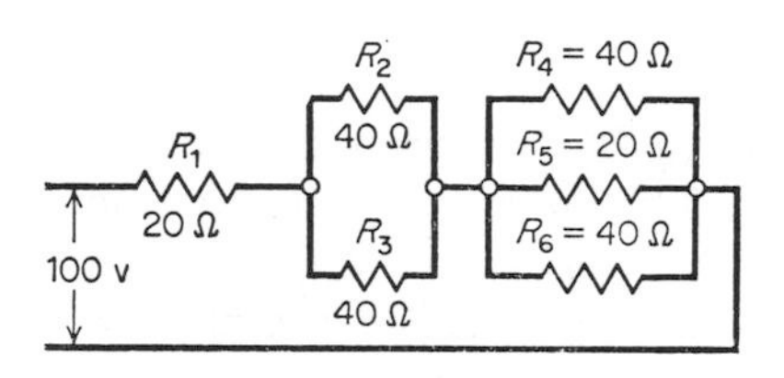

Figure 12.16

12.15. Calculate the full-load current drawn by a 440-volt, 20-hp induction motor having a full-load efficiency of 0.86 and a full-load power factor of 0.76.

Ans. 52 amps

12.16. A cubical framework is put together as shown in figure 12.17. Each member is identical and has an electrical resistance of "R." What is the equivalent resistance across the diagonal AB?

Ans. $5R/6$

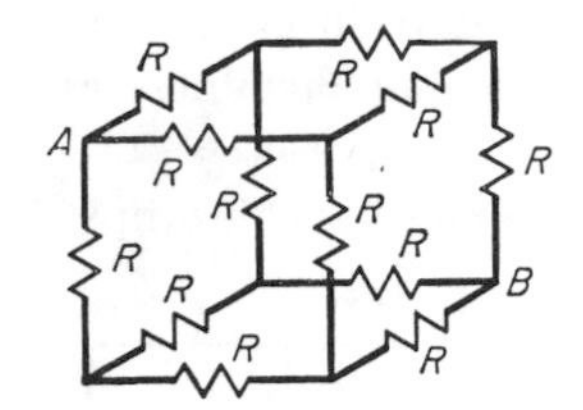

Figure 12.17

12.17. A 120-volt, 60-cycle, AC circuit consists of a capacitive reactance of 46 ohms and a resistance of 11 ohms connected in series. Determine (a) the impedance of the circuit, (b) the current, (c) the size inductor (in millihenrys) that must be connected in series so that maximum current will flow in the circuit and the value of the maximum effective current that will flow, (d) the combined voltage drop across the capacitor and inductor under the conditions of part (c).

Ans. (a) 47.3 ohms, (b) 2.54 amps, (c) 122 mh, 10.9 amps, (d) zero

12.18. An alternating voltage of 110 volts at 60 Hz is impressed on a series circuit of 8 ohms resistance, 0.0955 henrys inductance, and 42 ohms capacitive reactance. Compute (a) the impedance of the circuit in ohms, (b) the current through the circuit, (c) the power factor of the circuit, (d) the voltage drop across the inductor.

Ans. (a) 10 ohms, (b) 11 amps, (c) 0.8 leading, (d) 396 volts

12.19. When 10 volts DC is applied across a series circuit, the current flowing is 0.1 amp. When the DC supply is replaced by an AC supply, it is found that a voltage of 10.0 volts is necessary to cause a current of 0.05 amp to flow in the circuit. Find: (a) the DC resistance of the circuit, (b) the power absorbed under DC conditions, (c) the impedance of the circuit under AC conditions, (d) the power of the AC circuit.

Ans. (a) 100 ohms, (b) 1 watt, (c) 200 ohms, (d) 25 W

12.20. A small industrial plant operates with an average load of 400 kW, consisting of small induction motors with some electric heating. The average power factor is 60% lagging. A 150-kW synchronous motor driving a compressor is added to the plant. Neglecting the effect of motor losses, calculate the power factor at which this motor must be operated in order to raise the power factor of the plant to 80%.

Ans. 77.6%

12.21. (a) In the circuit shown in figure 12.18, give the amount of the in-phase current. (b) Give the line current. (c) Give the power absorbed by the circuit.

Ans. (a) 3.39 amps, (b) 3.42 amps, (c) 372 W

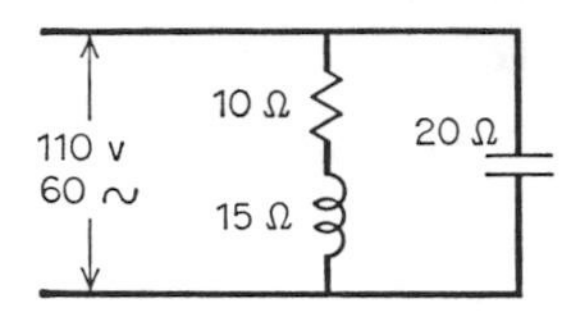

Figure 12.18

12.22. The primary of an r-f transformer has an inductance of 350 microhenrys. (a) What is its inductive reactance at 1200 kilocycles? (b) What current will flow when the voltage across the primary is 10 volts?

Ans. (a) 2640 ohms, (b) 3.79 milliamperes

12.23. A capacitance, inductance, and resistance in series are connected to an alternating current supply of 120 volts. Indicate true statement(s): (a) The arithmetic sum of the voltages across each piece of the equipment equals 120 volts. (b) The algebraic sum of the voltages across each piece of equipment equals 120 volts. (c) The vector sum of the voltage across each piece of equipment equals 120 volts. (d) The voltage across each piece of equipment equals $\sqrt{2}$ times the line voltage.

Ans. (c)

12.24. The impedance of a circuit is given as

$$Z = 5 + j10$$

(a) What is the magnitude of the resistance component? (b) What is the absolute magnitude of the impedance of the circuit?

Ans. (a) 5 ohms, (b) 11.2 ohms

12.25. The impedance of a series AC circuit is

$$Z = 4 + j3$$

The circuit consists of a resistor, capacitor, and inductor. (a) Which has the greater reactance, the capacitor or the inductor? (b) If a 120-volt, 60-cycle source is connected to the circuit, what is the magnitude of the current flowing?

Ans. (a) inductor, (b) 24 amps

12.26. The admittance of an alternating current circuit is given as

$$Y = 4 - j3$$

(a) What is the simplified complex expression for the impedance? (b) What is the magnitude of the pure resistance in the circuit?

Ans. (a) $0.16 + j0.12$, (b) 0.16 ohms

12.27. For the circuit shown in figure 12.19, what is the magnitude of (a) the impedance of the circuit, (b) the line current, (c) the voltage across AB, (d) the power factor of the circuit?

Ans. (a) 39.7 ohms, (b) 3.02 amps, (c) 401 volts, (d) 0.251 leading

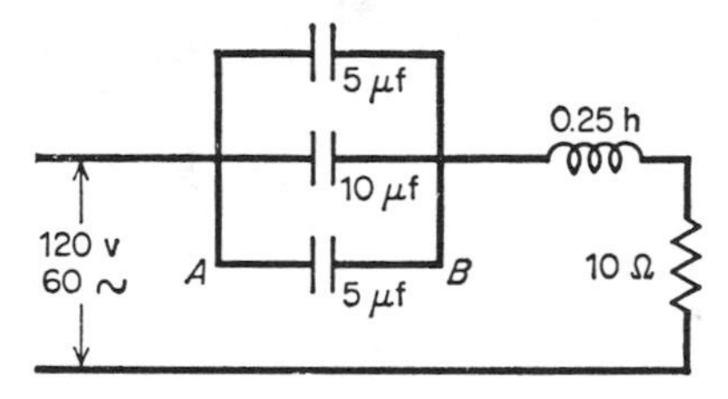

Figure 12.19

12.28. A 200-hp, 3-phase, 4-pole, 60-cycle, 440-volt, squirrel-cage induc tion motor operates at full load, with an efficiency of 85%, a power factor of 91%, and a slip of 3%. For this full-load condition, determine the following: (a) the speed in rpm, (b) the torque delivered in lbf-ft, (c) the linc current fed to the motor.

Ans. (a) 1746 rpm, (b) 602 lbf-ft, (c) 253 amps

12.29. In a vacuum tube, the electrons must leave a conductor and enter the vacuum. What causes the electrons to leave the conductor and enter the vacuum?

Ans. Electrons are agitated by the application of heat to the conductor, and are "boiled off" into space.

12.30. A power company delivers 500 kW to factory A, 250 kva at 0.85 power factor lagging to factory B, and 750 kva at 0.97 power factor leading to factory C. (a) Draw a vector diagram showing the loads, power factor, and total kva. (b) What is the total power load in kW for the three companies?

Ans. (a) pf = 0.99, 1441 kva, (b) 1440 kW

12.31. The voltage across the terminals of an AC circuit is given as

$$E = 4 + j3$$

What is the magnitude of (a) the inphase voltage, (b) the absolute voltage?

Ans. (a) 4 volts, (b) 5 volts

12.32. In the circuit shown in figure 12.20, what frequency of alternating current will cause the maximum current to flow with the rms voltage remaining constant?

Ans. 82.2 Hz

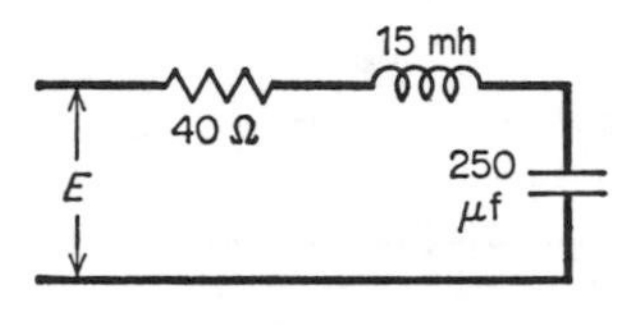

Figure 12.20

12.33. The following test data are available on a 50 kva, 2300/230-volt, 60-cycle distribution transformer.

open circuit—low side excited:

$E_s = 230$ volts
800 W
$I_s = 20$ amp

short circuit—high side excited:

$E_p = 60$ volts
1000 W
$I_p = 21.7$ amp

Find the percent impedance and percent resistance. Calculate the regulation and efficiency at 100% load, 80% lagging power factor.

Ans. 2.6%. 2%, 2.6%, 95.7%

12.34. A 15-hp, 230-volt, 60-cycle, 3-phase induction motor having an efficiency of 90% and a power factor of 85% is to be served from a panel board 200 feet away. If the voltage at the panel board is 235 volts, what is the minimum size of wire that can be used so that the voltage at the motor terminals will be 230 volts at full-load running conditions?

Ans. No. 8 AWG

12.35. Two 15-kva transformers are offered a purchaser. Transformer *A* costs $131.00, and has a core loss of 90 watts and a copper loss of 250 watts. Transformer *B* costs $108.00, and has a core loss of 95 watts and a copper loss of 255 watts. The transformer is to be operated 4 hours a day full-load, unity power factor, and 20 hours a day no-load, throughout the year. On a basis of a 20-year life, 6% interest, and energy at 3 cents per kW-hr, which transformer would you purchase?

Ans. B

12.36. All of the electrical energy supplied to an electric water heater is converted to heat. Due to the loss of heat through the walls of the heater, only 90% of the electrical energy supplied to a certain heater is used in raising the temperature of the water. With an applied voltage of 115 volts, the heater takes a current of 9 amp at unity power factor. (a) How long will it take to raise the temperature of a gallon of water from 20 °C to the boiling point? (b) At a cost of 5 cents per kW-hr, how much will the energy used in raising the temperature of the gallon of water cost?

Ans. (a) 22.7 minutes, (b) 1.95¢

12.37. An electrical milliammeter, having an internal resistance of 500 ohms, reads full scale when one milliampere (0.001 amp) of current is flowing. (a) If the meter is to be used as an ammeter reading 0.1 amp full scale, what size resistor (in ohms) should be used in parallel with it? (b) If used as a voltmeter reading 100 volts full scale, what size resistor (in ohms) should be used in series with the milliammeter?

Ans. (a) 5.05 ohms, (b) 99,500 ohms

12.38. (a) Battery *A* has a no-load terminal voltage of 9 volts and an internal resistance of 2 ohms. Battery *B* has a no-load terminal voltage of 6.5 volts with an internal resistance of 1 ohm. When the positive terminals of the two batteries are connected together, the negative terminals connected together, and a 3-ohm resistor connected between positive and negative terminals, what current will flow through each battery? (b) When a certain coil is connected to a 60-volt battery, 2 amp of current flow. When the same coil is connected to a 50-volt, 60-cycle AC source, only 1 amp flows. What is the inductance of the coil in henrys?

Ans. (a) $I_A = 1.5$ amps, $I_B = 0.5$ amps, (b) 0.106 h

12.39. The name plate of a 240-volt DC motor states that the full-load line current is 67 amp. On test at no load at rated voltage, it requires 3.35-amp armature current and 3.16-amp field current. Armature resistance at no load and at full load was 0.207 ohms. Brush drop was 2 volts. If stray power is assumed constant from no load to full load, what are the horsepower and efficiency at full load?

Ans. 18.2 hp, 84.5%

12.40. The load on a single-phase, 120-volt alternating-current generator consists of 200 incandescent lamps, each taking 1.0 amp at 120 volts. Certain small motors also are connected across the line, delivering a total of 40 hp with an average efficiency of 80% and an average power factor of 0.80. Find (a) the current output of the generator, (b) kilowatt output of the generator, (c) power factor of the total load, (d) horsepower of the driving engine if the generator efficiency is 90%.

Ans. (a) 512 amps, (b) 61.3 kW, (c) 0.91, (d) 91.3 hp

12.41. A resistance of 6 ohms and an unknown impedance coil in series takes 12 amp from a 120-volt, 60-cycle AC line. If the total power taken by the circuit is 1152 watts, what are the values of resistance and inductance of the coil?

Ans. 2 ohms, 6 ohms

12.42. An AC motor delivers 10 hp at 120 volts. The efficiency of the motor is 85%, and the power factor is 0.80 lagging. (a) What is the full load current? (b) What is the full load current of a 10-hp, 120-volt, DC motor of 85% efficiency?

Ans. (a) 91.3 amps, (b) 73.1 amps

12.43. A 250-hp, 440-volt, squirrel-cage induction motor operates at an efficiency of 90%, and a power factor of 85% at full load. What kva of capacitors is required to correct the power factor to unity at full load?

Ans. 128 kva

12.44. A non-inductive resistance and a condenser are connected in parallel. The parallel combination takes a current of 17 amp and absorbs 1400 watts from a 220-volt, 60-cycle source. (a) What are the currents in each branch? (b) What are the magnitudes of the components R and C?

Ans. (a) 6.36 amp, 15.76 amp, (b) 34.6 ohms, 190 μF

12.45. A balanced, Y-connected load of non-inductive resistances is connected in parallel with a balanced delta-connected load across a balanced 220-volt, three-phase circuit. The total power absorbed by both loads is 20 kW. If the impedance of each delta branch is $5 + j10$, what is the resistance of each Y branch? What will the ammeter read for total line current?

Ans. 3.4 ohms, 60.5 amps

12.46. A coil, whose impedance is given by the complex expression $2 + j8$, is connected in series with a non-inductive resistance of 6 ohms. What non-inductive resistance must be placed in parallel with the impedance coil so that the power factor of the entire circuit will be 0.8?

Ans. 38.5 ohms

12.47. A lamp load taking 100 amp and an induction motor are connected at the end of a transmission line whose total impedance is $0.1+j0.2$. When starting, this motor takes 10 kW at 0.4 power factor. If the potential at the load is to be kept 230 volts, what must be the impressed potential at the other end of the line when the motor is starting?

Ans. $264.3 + j18.7$ volts

12.48. An induction motor takes 47.2 amp from a 220-volt, 60-cycle source when running full-load, and delivers 10 hp at an efficiency of 85%. What value of capacitance must be connected in series with the motor in order that the power factor of the system be unity when there is a full load on the motor? (Assume the same efficiency and power factor for the motor.)

Ans. 1060 μF

12.49. Two impedances, $Z_1 = 10 + j10$ and $Z_2 = 0 - j10$, are connected in parallel, and the combination is connected in series with a third impedance which has a resistance of 10 ohms and a variable reactance. (a) What value of the unknown reactance makes the power taken by the entire circuit a maximum? (b) What is the maximum power if the applied voltage is 300 volts? (c) Is the unknown reactance capacitive or inductive?

Ans. (a) 10 ohms, (b) 4500 W, (c) inductive

12.50. How many amperes does a 100-watt light bulb take when running on a 110-volt line?

Ans. 0.91

12.51. Three resistors, A, B, and C, are connected in parallel across an

80-volt line. The resistance of each, in ohms, is as follows: $A = 2$; $B = 4$; $C = 8$. What current will flow?

Ans. 70 amps

12.52. A 220-volt, AC motor is to be connected to a 4400-volt transmission line. The device used for this purpose is: (a) a 20-to-1 capacitor, (b) a 5-to-1 transmitter, (c) a 50-to-1 regulator, (d) a 2-to-1 transformer, (e) a 20-to-1 transformer.

Ans. (e)

12.53. A hoist is to be powered with an electric motor, and both DC and AC lines are available. The best type of motor for this hoist is: (a) DC series, (b) AC split-phase, (c) DC shunt-wound, (d) AC series, (e) DC dynamometer.

Ans. (a)

12.54. Power can be transmitted through a certain AC transmission line at a voltage of either 100 volts or 10,000 volts. To transmit a fixed number of kilowatts, the higher voltage is more efficient by a factor of: (a) 10,000 to 1, (b) 1000 to 1, (c) 100 to 1, (d) 10 to 1, (e) 500 to 1.

Ans. (a)

12.55. A certain machine must be run at precise and constant speed. Both DC and AC lines are available. The simplest way to furnish about $\frac{1}{30}$ hp to this machine is to use: (a) a DC shunt-wound motor, (b) an AC synchronous motor, (c) a DC series motor, (d) a DC motor with electron tube regulators, (e) an AC induction motor.

Ans. (b)

12.56. Alternating current is used rather than direct current because it is (a) easier to transmit, (b) better for traction motors, (c) easier to generate, (d) possible to transmit at high voltage and to utilize at low voltage, (e) less likely to interfere with television programs.

Ans. (d)

12.57. A coil has a reactance of 37.7 ohms and a resistance of 12 ohms. It is connected to a 110-volt, 60-cycle line. Determine: (a) the impedance of the coil, (b) current through the coil, (c) phase angle with respect to the supply voltage as a reference, (d) power factor of the circuit, (e) reading of a wattmeter connected to the circuit.

Ans. (a) 39.5 ohms, (b) 2.78 amps, (c) 72.3°, (d) 0.30 lagging, (e) 92 W

12.58. Two condensers are connected in series across as 12-volt DC source. One has a capacitance of 3 μF and the other has a capacitance of 6 μF. Calculate the charge for each condenser and the potential difference across each one.

Ans. 24 microcoulombs each, 8 volts, 4 volts

12.59. A battery consists of 6 cells, each having an *EMF* of 1.1 volts and a resistance of 4 ohms, connected in two parallel rows of three cells each in series. This battery is connected to a group of two resistances in parallel of 30 ohms and 6 ohms, respectively. The lead wire resistance is negligible. Using Ohm's law, calculate the current in the leads.

Ans. 0.3 amps

12.60. A single-phase transmission line 12 miles long has a conductor with a resistance of 0.27 ohms per mile and a reactance of 0.62 ohms per mile. The voltage at the load end is 13,200 volts. (a) Calculate the voltage required at the power house when the load is 1500 kva at unity power factor. (b) Recalculate with a power factor of 0.8 lagging. (c) Recalculate with a power factor of 0.8 leading.

Ans. (a) 13,570 volts, (b) 14,000 volts, (c) 13,000 volts

12.61. A three-phase motor load totaling 325 kW at 0.75 power factor is carried by three transformers, each rated at 150 kva, and connected in delta. (a) If one transformer is disconnected, calculate the percent of overload on the remaining two. (b) What load in kW at 0.75 power factor could two of these transformers carry without exceeding their rated load?

Ans. (a) 66.7%, (b) 195 kW

12.62. A 250-volt, 10-hp, DC shunt motor has an efficiency of 85%. Draw a wiring diagram that shows how a starting resistor should be connected. Specify the ohmic value of a starting resistor that limits the starting current to half the full-load value. Estimate the starting torque in percent of full-load value.

Ans. 14.3 ohms, 50%

13 *CHEMISTRY*

SYMBOLS

B = volumetric fraction or percentage of one component of a mixture of gases
c = specific heat
d = rate of diffusion
e = base of natural logarithms
G = gravimetric fraction or percentage of one component of a mixture
M = molecular weight
m = mass
p = pressure
T = absolute temperature
t = time
V = volume
δ = decay constant
ρ = density

Atomic Weights

The following are atomic weights for a few substances, based on oxygen being exactly 16:

aluminum	Al	26.97	lead	Pb	207.21
argon	A	39.94	magnesium	Mg	24.32
boron	B	10.82	mercury	Hg	200.61
calcium	Ca	40.08	nitrogen	N	14.008
carbon	C	12.01	oxygen	O	16.000
chlorine	Cl	35.46	phosphorus	P	30.98
copper	Cu	63.54	radium	Ra	226.05
helium	He	4.003	silicon	Si	28.06
hydrogen	H	1.008	sodium	Na	22.997
iron	Fe	55.85	sulfur	S	32.066

Selected Molecular Weights

acetylene	C_2H_2	26.04	hydrogen sulfide	H_2S	34.08
air (equivalent)		28.97	isobutane	C_4H_{10}	58.12
ammonia	NH_3	17.03	methane	CH_4	16.04
chlorine	Cl_2	70.91	nitric oxide	NO	30.008
carbon dioxide	CO_2	44.01	nitrogen	N_2	28.02
carbon monoxide	CO	28.01	nitrous oxide	N_2O	44.02
ethane	C_2H_6	30.07	oxygen	O_2	32.000
ethylene	C_2H_4	28.05	propane	C_3H_8	44.09
hydrogen	H_2	2.016	sulfur dioxide	SO_2	64.07
hydrogen chloride	HCl	36.47	octane	C_8H_{18}	114.224

BASIC LAWS

Questions on the examinations relating to chemistry often have to do with combustion processes. Such questions are sometimes under the heading of thermodynamics. Laws already given that overlap chemistry and thermodynamics (physics) include Boyle's law, Charles' law, Archimedes' law, Dalton's law, and conservation of mass and energy.

Avogadro's law states that equal volumes of all ideal gases at a particular pressure and temperature contain the same number of molecules. The number of molecules in a gram-mole is of the order of 6 EE23, depending on its state.

Graham's law states that at a particular temperature and pressure, the rates of diffusion of two gases are inversely proportional to the square roots of their densities:

$$d_1/d_2 = \sqrt{\rho_2/\rho_1}$$

Gay-Lussac's law states that the combining volumes of reacting gases and the resulting volumes of the products are always in simple whole-number proportions when measured at the same pressure and temperature.

Dulong and Petit discovered that, for elements in the solid state, the product of the atomic weight and the specific heat is the same for all elements, such that c (in BTU/lbm-°F or cal/g-°C) times atomic weight approximately equals 6.4. Specific heats vary with temperature, and the law therefore involves an approximation, but, as the temperature increases, the product approaches 6.4, even when it is quite different at room temperatures.

VALENCE

With exceptions, the valence of hydrogen and of sodium is +1; chlorine, −1; oxygen, −2 (except in peroxides, −1); magnesium, +2; aluminum, +3; nitrogen, −3; carbon, −4; and so forth. The valence of an element or radical when it is part of a compound is the number of atoms of hydrogen with which it combines, or the number of atoms of hydrogen it can displace (lowest-common-multiple compounds). Metals have positive valence, and nonmetals have negative valence. The valence of an uncombined element or of a compound, such as Al_2O_3, is zero. This knowledge is used to help decide what compounds are formed by a chemical reaction. For example, if Al_2O_3 is to combine with H_2SO_4, the aluminum does not combine with the hydrogen in H_2SO_4 because they both have positive valence. Since the SO_4 has −2 valence, Al and SO_4 are likely to combine. In order for their combination to have zero valence (Al with +3), use 2 atoms of Al and 3 SO_4 radicals, giving $+6 - 6 = 0$:

$$Al_2O_3 + 3H_2SO_4 \rightarrow Al_2(SO_4)_3 + 3H_2O$$

This equation also has been balanced for mass (same number of atoms of each element on each side of the equation).

The combination SO_4^{-2} is called a *radical* and has a valence of −2. Other common radicals and their valences are NO_3^{-1}, PO_4^{-3}, OH^{-1}, and NH_4^{+1}.

The coefficients of the chemical symbols are the number of said molecules involved, such as 3 molecules of H_2SO_4 above, and they are also the amount of each substance in moles. M is molecular weight (M lbm/mole or M g/mole for a pound mole and gram mole, respectively). If all substances are ideal gases (not true in the foregoing reaction), the volumes involved are proportional to the coefficients when all gases are at the same pressure and temperature (Avogadro's law). The actual masses involved in the reaction are proportional to the molecular weights times the coefficients.

GRAVIMETRIC AND VOLUMETRIC ANALYSES

A *gravimetric analysis* (G_x%, G_y%, of substances X and Y) gives the percentage (or fraction) by weight of each substance in a mixture. A *volumetric analysis* (B_x%, B_y% of gases X and Y, used for gases only) gives the percentage (or fraction) by volume of each gas in a mixture. To convert from gravimetric analysis to volumetric, say X, Y, and Z, divide each G, lbm/lbm, by the molecular weight, M lbm/mole, and sum these terms.

$$\frac{G_x}{M_x} + \frac{G_y}{M_y} + \frac{G_z}{M_z} = \sum \frac{G}{M} \frac{\text{mole}}{\text{lbm}}$$

You alternatively can use mole per other mass unit: Divide each individual ratio by $\sum \frac{G}{M}$.

$$\frac{G_x/M_x}{\sum G/M} = B_x \frac{\text{mole}}{\text{mole}}$$

This gives the volumetric percentage or fraction. To convert from volumetric to gravimetric, multiply each B, mole/mole, by M, lbm/mole, and sum these terms.

$$B_x M_x + B_y M_y + B_z M_z = \sum BM \text{ lbm/mole}$$

Divide each individual product by $\sum BM$.

$$\frac{B_x M_x}{\sum BM} = G_x \frac{\text{lbm}}{\text{mole}} \text{ of mixture}$$

The mole is also a measure of volume for gases (Avogadro).

At standard conditions of 32 °F (0 °C) and 14.696 psi (760 mm Hg), a pound-mole of an ideal gas occupies 359 cu ft. (A gram-mole occupies 22.4 liters or 0.7907 cu ft.)

MISCELLANEOUS RULES

Knowledge of the mole volume permits the calculation of the approximate molecular weight. For example, if 3.64 l of some gas at 100 °C and 913 mm Hg weighs 2.43 g at standard gravity, its molecular weight is found as follows. Convert the volume to that at standard conditions from $p_1V_1/T_1 = p_sV_s/T_s$, or

$$V_s = V_1\left(\frac{p_1}{p_s}\right)\left(\frac{T_s}{T_1}\right) = 3.64\left(\frac{913}{760}\right)\left(\frac{273}{373}\right) = 3.2\ l$$

The density at standard conditions is then

$$\rho_s = 2.43/3.2 = 0.76\ \text{g}/l$$

The mass of 22.4 l is $(22.4)(0.76) = 17$, the approximate molecular weight.

The partial pressure of a single gas, X, in a mixture of ideal gases is given by $p_x = B_x p_m$, where p_m is the total pressure of the mixture.

The volume of any component in a gaseous mixture is the volume of the mixture.

The density of a mixture is the sum of the densities of the components ($\rho_m = \rho_x + \rho_y + \rho_z$ for gases X, Y, and Z).

The number of moles of each gas in a mixture is propotional to its volumetric fraction or percentage.

The symbol C_3H_8(g) means that the substance, propane, is in gaseous form: C_3H_8(l) is liquid. When the heat of reaction (heating value in combustion problems) is involved, the foregoing distinction is necessary for reactants and products whenever a substance might be in either phase, because it makes a difference in the energy released.

The symbol of the molecule reveals the gravimetric percentages of the atoms. M for C_3H_8 is approximately $(3)(12) + 8 = 44$. Then, divide 36 by 44 to get 0.818 or 81.8% C and 8 by 44 to get 18.2% H_2 by weight, the gravimetric composition of propane.

Conversely, for 80% C and 20% H_2, gravimetric, there are $80/12 = 6.67$ moles C and $20/1 = 20$ moles of H_2. These numbers are proportional to the number of atoms; that is, the formula could be $C_{6.67}H_{20}$, C_2H_6, or

CH_3. This process might give only a hypothetical average molecule of a mixture, as in ordinary hydrocarbon products such as gasoline, or it would likely be assumed as the most stable form, if it were known to be a single substance.

SOLUTIONS

When one substance goes into *solution* with another, the dispersion of the solute is molecular; that is, no finite particles of the dissolved substance exist. In *suspensions*, one substance disperses in the other in particles (not single molecules), in some cases small enough to be invisible to ordinary microscopic observation. A mixture of immiscible liquids is an *emulsion.*

In a *normal solution*, the amount of solute in 1 liter of solution is equal to the *equivalent weight* in grams. The equivalent weight of an element is the number of grams of a substance or compound that combines with one (1.008) gram of hydrogen or with 8 g of O_2.

In a *molal solution*, there is one gram-molecular weight of solute in 1000 g of solvent. *Molality* means the number of gram-molecular weights of solute in 1000 g of solvent. For example, the molality of a solution that contains 49.04 g of sulfuric acid ($M = 98.08$) dissolved in 1000 g of water is 0.5. A molal solution in 1000 g H_2O has $98.08 + 1000 = 1098.08$ total grams.

In a *molar solution*, there is one gram-molecular weight of the solute in one liter of solution. *Molarity* means the number of gram-molecular weights per liter of solution.

Ions in Solutions

When a substance goes into solution, some of its molecules dissociate or ionize into positive and negative ions, called respectively *cations* and *anions.* A strong electrolyte is one that completely dissociates (ionizes) in aqueous solutions; for example,

$$\mathrm{NaCl} \rightarrow \mathrm{Na^+} + \mathrm{Cl^-}$$

On the other hand, acetic acid only partially ionizes.

$$\mathrm{HC_2H_3O_2} \rightleftharpoons \mathrm{H^+} + \mathrm{C_2H_3O_2^-}$$

For 1 gram-mole in 16 liters, less than 2% will be ionized.

Cations and anions are indicated by superscript plus and minus signs, one for each valence, such as Ca^{++}, Al^{+++}, Cl^{-}, SO_4^{--}, and PO_4^{---}. The solution is electrically neutral, equal numbers of cations and anions. Chemical properties of ions are different from those of electrically neutral atoms of the same substance.

The pH value is the logarithm to the base 10 of the reciprocal of the H^+ ion concentration in g/l, a value that ranges from 0 to 14. A pH of 7 is neutral.

$$\text{pH} > 7 \quad (\text{OH}^- \text{ exceeds H}^+) \text{ is alkaline}$$

$$\text{pH} < 7 \quad (\text{H}^+ \text{ exceeds OH}^-) \text{ is acid}$$

Water hardness and scale-forming is often principally due to calcium bicarbonate, $Ca(HCO_3)_2$, and magnesium bicarbonate, $Mg(HCO_3)_2$. If the water containing these bicarbonates is boiled, they react to form the relatively insoluble $CaCO_3$ and $MgCO_3$ that, given time, settle out and make the water softer. Scale-forming impurities not so easily removed are calcium sulfate, $CaSO_4$, and magnesium sulfate, $MgSO_4$. Both calcium hydroxide, $Ca(OH)_2$, and sodium carbonate, Na_2CO_3, are used for softening, as well as the zeolites ($NaAlSiO_4$).

COMBUSTION

With exceptions, such as some exotic fuels, combustion usually occurs in atmospheric air (approximately 21% O_2 and 79% N_2 by volume), in which case, use $79/21 = 3.76$ moles N_2/mole O_2. The whole number nearest to the exact atomic weight is accurate enough for processes whose heat of reaction is desired.

With usual fuels (petroleum derivatives, coal, and so forth), the products of complete combustion are CO_2, H_2O, and SO_2 if the fuel contains sulfur (often ignored in computations). If the combustion is incomplete, there are also CO, CH_4, and H_2. In any case, inert gases pass through the process unchanged and excess O_2 is assumed to be O_2 after the reaction.

The method of balancing combustion equations and the information obtained therefrom is illustrative of chemical equations in general, and examples give the best review of the use of the foregoing principles.

EXAMPLE 1: A gaseous fuel is composed of 60% H_2 and 40% CH_4 by volume. Determine (a) the pounds of air per pound of fuel for complete ideal combustion, (b) the mass of H_2O formed per pound of fuel, (c) the volumetric analysis of the products, H_2O not condensed, (d) the gravimetric analysis of the reactants, (e) the partial pressure of the H_2O vapor for a total pressure of 14.7 psia.

SOLUTION: (a) Get the chemical equation first and then show the various relative masses and volumes. If the unknown amount of O_2 needed is aO_2, then the atmospheric N_2 is $3.76aN_2$. The products of combustion are CO_2 and H_2O. Then, for 1 mole of fuel,

$$0.6H_2 + 0.4CH_4 + aO_2 + 3.76aN_2 \rightarrow bCO_2 + cH_2O + 3.76aN_2$$

Mass balances are:

$$\text{for } O_2, \quad 2a = 2b + c$$

$$\text{for } H_2, \quad 1.2 + 1.6 = 2c$$

$$\text{for C}, \quad 0.4 = b$$

From these, we get $b = 0.4$, $c = 1.4$, and $a = 1.1$. For 1 mole of fuel $(0.6 + 0.4)$, the chemical equation becomes

	$0.6H_2$	$+ 0.4CH_2$	$+ 1.1O_2$	$+ 4.14N_2$	$\rightarrow 0.4CO_2$	$+ 1.4H_2O$	$+ 4.14N_2$
R.V.	0.6	+ 0.4	+ 1.1	+ 4.14	→ 0.4	+ 1.4	+ 4.14
R.M.	1.2	+ 6.4	+ 35.2	+ 115.9	= 17.6	+ 25.2	+ 115.9
$\frac{\text{mass}}{\text{lbm fuel}}$	1		+ 4.63	+ 15.25	= 2.31	+ 3.32	+ 15.25

R.V. is relative volumes, which are the coefficients of gaseous substances. (If a solid or liquid had been involved, its volume would have been taken as zero.) R.M. is relative masses. The R.M. is the product of the molecular weight and the coefficient (moles). The mass of fuel is $1.2 + 6.4 = 7.6$ lbm/mole (or g/mole). The fourth line, mass per pound of fuel, is obtained by dividing relative weights by 7.6. Since the air is the sum of O_2 and N_2,

$$4.63 + 15.25 = 19.88 \text{ lbm air/lbm fuel} \quad (\textit{Ans.})$$

(b) The mass of H_2O is 3.32 lbm/lbm fuel. (*Ans.*)

(c) The total volume of gaseous products is

$$0.4 + 1.4 + 4.14 = 5.94 \text{ moles (say)}$$

$$B_{CO_2} = 0.4/5.94 = 6.74\%$$

$$B_{H_2O} = 1.4/5.94 = 23.58\%$$

$$B_{N_2} = 4.14/5.94 = 69.7\%$$

(d) The total mass of reactants is

$$1.2 + 6.4 + 35.2 + 115.9 = 158.7 \text{ (any mass unit)}$$

$$G_{H_2} = 1.2/158.7 = 0.76\%$$

$$G_{CH_4} = 6.4/158.7 = 4.03\%$$

$$G_{O_2} = 35.2/158.7 = 22.18\%$$

$$G_{N_2} = 115.9/158.7 = 73.03\% \quad (\textit{Ans.})$$

(e) The partial pressure of the H_2O is

$$p = Bp_m = (0.2358)(14.7) = 3.46 \text{ psia} \quad (\textit{Ans.})$$

OXIDATION AND REDUCTION

Oxidation and reduction do not always involve oxygen. *Oxidation* is a process in which an element loses electrons; that is, the valence number increases and it becomes more positive. *Reduction* is a process in which an element gains electrons; that is, decreases in valence and becomes less positive. Oxidation and reduction occur simultaneously and the decrease of electrons is equal to the gain of electrons.

$$\overset{+4}{MnO_2} + \underset{-4}{4HCl} \rightarrow \overset{+2}{MnCl_2} + \underset{0}{Cl_2} + 2H_2O$$

In the above reaction, Mn in MnO_2 has a valence of +4 (because O_2 is −4). In $MnCl_2$, the valence of Mn is +2 (Cl is −1). Therefore, Mn, having gained 2 electrons (lost valence), is reduced. The Cl_2 in 4HCl has a valence of −4 (H is +1), and the valence of Cl_2 alone is zero. Therefore, Cl_2, having lost 2 electrons (gained valence), is oxidized. A change in the valence number of an atom or radical must be balanced by an opposite change of another.

ATOMIC DISINTEGRATION

The equation of decay of a radioactive substance is

$$m = m_o e^{-\delta t}$$

m is the mass of this substance remaining after time t, its decay (or disintegration) constant is δ, m_o is the original mass, and e is natural logarithm base. The half-life is found by letting $m = m_o/2$; this gives (taking natural logarithms)

$$t = \frac{\ln 2}{\delta} = \frac{0.693}{\delta} \qquad \text{[HALF-LIFE]}$$

This is the time for half of the radioactive substance existing at any instant to decay.

PRACTICE PROBLEMS

13.1. Balance the following chemical equation. How many moles of hydrogen are liberated per mole of aluminum?

$$\_Al + \_H_2SO_4 \rightarrow \_Al_2(SO_4)_3 + \_H_2$$

13.2. Balance the following chemical equation and name the elements and radicals represented by the chemical symbols:

$$\_Ag + \_O_2 + \_NaCN + \_H_2O \rightarrow \_NaAg(CN)_2 + \_NaOH$$

13.3. Hydrogen sulfide gas, H_2S, is used to precipitate lead sulfide, PbS, in a solution of hydrochloric acid.

$$\_H_2S + \_PbCl_2 \rightarrow \_PbS + \_HCl$$

(a) Balance the chemical equation and state how many moles of H_2S are required to produce one mole of PbS. (b) How many pounds of H_2S are required to produce one pound of PbS? (c) How many cubic feet of H_2S gas at 70 °F and atmospheric pressure are required to produce one pound of PbS? The gas constant for H_2S is 45.33 ft-lbf/lbm-°R.

Ans. (a) 1, (b) 0.143 lbm, (c) 0.845 ft^3

13.4. Potassium chlorate ($KClO_3$) will liberate all of its oxygen if heated to the proper temperature. Potassium nitrate (KNO_3) will liberate part of its oxygen if heated to the proper temperature, yielding $KNO_2 + O_2$. What weight of KNO_3 is required to deliver the same amount of oxygen as one pound of $KClO_3$?

Ans. 2.47 lbm

13.5. Complete and balance the following chemical reactions by filling in the blanks:

(a) $\_CH_4 + \_Cl_2 \rightarrow \_C + \_\_$

(b) $\_AgNO_3 + \_HCl \rightarrow \_\_ + \_HNO_3$

(c) $\_AsCl_3 + \_H_2S \rightarrow \_As_2S_3 + \_HCl$

(d) $\_Cu_2O + \_Cu_2S \rightarrow \_Cu + \_SO_2$

(e) $\_B_2O_3 + \_Mg \rightarrow \_MgO + \_B$

(f) $\_BaSO_4 + \_C \rightarrow \_BaS + \_CO$

13.6. The half-life of a radioactive element is the time required for half of a given quantity of that element to disintegrate into a new element. For example, it takes 1600 years for half of a given quantity of radium to change to radon. In another 1600 years, half the remainder will have disintegrated, leaving one-fourth of the original amount. The half-life of radium is therefore said to be 1600 years. If the rate of disintegration of another radioactive material is 35 parts per 100 every hour, what is its half-life?

Ans. 1.98 hours

13.7. Pure silver (Ag) is combined with nitric acid (HNO_3) to yield silver nitrate ($AgNO_3$), nitric oxide (NO), and water (H_2O). (a) Write the balanced chemical equation for this reaction. (b) How many pounds of silver would be required to liberate 100 cu ft of nitric oxide at 60 °F and one atmosphere pressure? The gas constant for nitric oxide is 49.5 ft-lbf/lbm-°F.

Ans. (b) 88.7

13.8. Propane, C_3H_8, is completely burned in air with carbon dioxide, CO_2, and water, H_2O, being formed. (a) Write the balanced chemical equation. (b) If 15 pounds of propane are burned per hour, how many cu ft/hr of dry CO_2 is formed after it has cooled to 70 °F and atmospheric pressure? The gas constant for CO_2 is 35.0 ft-lbf/lbm-°R.

Ans. (b) 394 ft^3

13.9. According to Avogadro's law, equal volumes of all gases at the same temperature and pressure contain the same number of molecules. Furthermore, 22.4 liters of a gas contain 6.06 EE23 molecules at 0 °C and one atmosphere of pressure. How many molecules of nitrogen, N_2, are contained in 10 liters of nitrogen saturated with water vapor at 60 °C and under a total absolute pressure of two atmospheres? (Absolute zero can be taken as −273 °C and the vapor pressure for water is 3.1 psia at 60 °C.)

Ans. 3.97 EE23

13.10. A reversible chemical reaction is one that: (a) yields products that can react among themselves to give back the original reactants, (b) yields several products, any of which must be soluble in water, (c) requires a catalyst to complete the reaction, (d) will remain in equilibrium, regardless of the changes in heat, pressure or concentration of the reactants without affecting the amount of the products formed, (e) must have either the reactants or products in a gaseous state.

Ans. (a)

13.11. Balance the following chemical equations and state whether the products formed are precipitate, gas, or in solution.

$$\_SnCl_2 + \_H_2S \rightarrow \_SnS + \_HCl$$
$$\_Ag_2SO_4 + \_Cu \rightarrow \_Ag + \_CuSO_4$$
$$\_HI + \_H_2SO_4 \rightarrow \_H_2O + \_I_2 + \_H_2S$$

13.12. Carbon in living matter contains a definite proportion of ratio-carbon C^{14}. Upon death of this living matter, the radiocarbon decreases at the rate of one part in 8000 per year. (a) Derive an equation where A is the original amount of C^{14} at the death of the living matter and y is the amount of C^{14} left at any time t in years measured from the death of the living matter. (b) Charcoal from a tree killed by the eruption of the volcano that formed Crater Lake assayed at 44.5% radiocarbon. How many years ago did the eruption occur?

Ans. (a) $y = Ae^{-kt}$, (b) 6480 years

13.13. If the valence of phosphorus is 5, its oxide has the formula: (a) PO_5, (b) P_5O, (c) P_2O_5, (d) P_5O_2.

Ans. (c)

13.14. "Temporary hardness" of water is due to the presence of: (a) $Ca(HCO_3)_2$, (b) $CaSO_4$, (c) NaCl, (d) $NaCO_3$.

Ans. (a)

13.15. Air is approximately 20% O_2 and 80% N_2 by volume. If the air is at atmospheric pressure at sea level, what is the partial pressure of the oxygen?

Ans. 2.94 psia

13.16. When a pound-mole of methane is burned completely, how many pounds of O_2 are required?

Ans. 64 lbm

13.17. Hardness in water due to sulfates being present is reduced by: (a) trickling filters, (b) boiling, (c) a chemical reaction, (d) treating with sodium chloride.

Ans. (c)

13.18. The valence of phosphorus in the compound Na_3PO_4 is: (a) −5, (b) +5, (c) −3, (d) +3.

Ans. (b)

13.19. Which of the following processes is a physical change: (a) rusting, (b) combustion, (c) condensation of water vapor, (d) wood decay?

Ans. (c)

13.20. A compound, when analyzed, is found to contain 79.9% copper (at. wt 63.57) and 20.1% oxygen (at. wt 16) by weight. The simplest molecular formula for the compound is: (a) CuO, (b) Cu_2O, (c) CuO_2, (d) Cu_2O_3.

Ans. (a)

13.21. Methane (CH_4), when burned in air, will form water and carbon dioxide. Find the number of pounds of O_2 required for the complete combustion of 1 pound CH_4, the number of pounds of CO_2 formed and the number of pounds of H_2O formed.

Ans. 4 lbm, 2.75 lbm, 2.25 lbm

13.22. The specific gravity of a 10% calcium chloride solution ($CaCl_2$ dissolved in water) is 1.0835. Water weighs 62.43 lbm/cu ft at 39 °F and there are 7.48 gallons/cu ft. Calculate the number of pounds of calcium chloride required to make 55 gallons of a 10% solution of this salt at 39 °F.

Ans. 49.7 lbm

13.23. (a) Calculate the weight of lime, CaO, that can be prepared by heating 200 pounds of limestone containing 85% by weight of $CaCO_3$ and 15% by weight of inert material. (b) Calculate the weight of the mixture of lime, CaO, and the inert material after heating.

Ans. (a) 95.2 lbm, (b) 125.2 lbm

13.24. One of the principal scale-forming constituents of water is calcium carbonate, $Ca(HCO_3)_2$. Explain how this substance can be removed from water (a) by boiling, (b) by adding calcium hydroxide, $Ca(OH)_2$. Write the chemical equations.

13.25. Discuss the corrosion of iron. Include the following: (a) the chemical reaction that takes place, (b) factors that favor corrosion or increase the rate of corrosion, (c) methods of reducing the tendency toward corrosion. For each method cited, give reasons for its efficacy as best you can.

13.26. A gas analysis by volume gives the following: 12% CO_2, 4% H_2, 5% CH_4, 23% CO, and 56% N_2. (a) Find the percentages by weight. (b) Write the combustion equation. (c) How much air is required for complete combustion?

Ans. (a) 18.65%, 0.29%, 2.8%, 22.72%, 55.5%, (c) 1.24 lbm/lbm mixture

13.27. (a) A chemical classification of matter would put all material into one of two categories, pure substances and mixtures. What is the fundamental distinction between the two? Give examples of each. (b) What is a solution? Sometimes a solution is placed into one of the above categories, sometimes into the other. Explain. (c) The solid form of two elements are mixed and heated. As a result, you might obtain either a compound of the two, a solution, or a mixture. Give an illustration of each of these conditions and explain.

13.28. What weight of hydrogen peroxide, H_2O_2, could be prepared from 100 pounds of barium peroxide, BaO_2, by treating it with sulfuric acid, H_2SO_4, assuming that none of the hydrogen peroxide is lost?

Ans. 20.08 lbm

13.29. How many grams of sodium hydroxide are contained in one liter of a one-tenth normal solution?

Ans. 4 g

13.30. A sample of iron ore, FeO, weighing 0.3 g was dissolved in sulfuric acid. The resulting ferrous sulfate, $FeSO_4$, was oxidized to ferric sulfate, $Fe_2(SO_4)_3$, by potassium permanganate, $KMnO_4$. The oxidation required 28 cc of a solution of the permanganate that had been prepared by dissolving 3.16 g of the permanganate in 1 liter of water. Calculate the percentage of iron that was present in the ore.

Ans. 52%

13.31. Octane, C_8H_{18}, was combined with air and burned in a closed chamber, producing the following gas products analysis: 13% CO_2, 1% CO, 4% O_2, and 82% N_2, by volume. Find the original air-fuel ratio.

Ans. 14.9

13.32. A sample of copper weighing 10 g was heated with sulfuric acid. After the copper was dissolved, water was added. The solution was evaporated to dryness and 30 g of $CuSO_4 \cdot 5H_2O$ was obtained. Calculate the percentage purity of the sample of copper.

Ans. 76.5%

13.33. How many kilograms of iron will be required to produce 1000 liters of hydrogen at 80 °C and 780 mm pressure? The reaction is

$$3Fe + 4H_2O \rightarrow Fe_3O_4 + 4H_2$$

Ans. 1.49 kg

13.34. A volume of 287.5 ml of a vapor, measured at 100 °C and 752 mm pressure, weighed 0.725 g. What is the molecular weight of the compound?

Ans. 78.4 g/mole

13.35. An empty steel cylinder has a capacity of 2 cu ft. Its weight empty, with valve, is 135 pounds. The cylinder is filled with oxygen at 1900 psi and 77 °F. Assume the validity of the gas laws. A pound-mole of a gas at standard conditions occupies approximately 358 cu ft. (a) Calculate the number of cubic feet of atmospheric air at 77 °F required to fill the tank under the given conditions. (b) Calculate what percentage of the total weight of the full tank is oxygen.

Ans. (a) 260 ft^3, (b) 13.5%

13.36. Phosphorus is prepared in an electric furnace according to the chemical equation:

$$Ca_3(PO_4)_2 + 3SiO_2 + 5C \rightarrow 3CaSiO_3 + 5CO + 2P$$

Calculate (a) the number of pounds of phosphorus formed for each pound $Ca_3(PO_4)_2$ used, (b) the number of grams of SiO_2 required for each grain of P produced.

Ans. (a) 0.2 lbm, (b) 0.19 g

13.37. The complete combustion of propane gas is represented by the following skeleton equation:

$$C_3H_8 + O_2 = CO_2 + H_2O$$

(a) Balance the equation. (b) How many cu ft of air measured at 25 °C and 760 mm Hg would be required to burn 10,000 cu ft of propane measured at the same p and T?

Ans. (b) 238,000 ft^3

13.38. A current of 2.5 amp is passed through a solution of sulfuric acid for 1 hour. The hydrogen and oxygen liberated at the electrodes are collected as a mixed gas over water. What is the volume of the mixed gases collected at 25 °C and 780 mm Hg pressure, measured over water?

Ans. 1.71 l

13.39. The lightest commonly used metal for engineering purposes is which of the following: (a) aluminum, (b) gallium, (c) barium, (d) magnesium?

Ans. (d)

14 *PHYSICS*

SYMBOLS

E = luminous flux per unit area; modulus of elasticity
F = luminous flux, lumens; force
f = focal length; frequency
I = candles
i = angle of incidence
L = length
m = mass
r = radius, distance; angle of refraction
T = period of a vibration; $T = 1/f$
v = speed
λ = wave length
μ = refractive index
ρ = density

ILLUMINATION

The magnitude of a source of light is measured in standard *candles*, I. Luminous intensity, E, is measured in *foot-candles* (and lumens per square foot). Quantity of light, F, or luminous flux, is measured in *lumens*. A 1-candle *point* source projects one lumen on *unit* area at a *unit* distance (spherical) from the source, and the intensity of light on the area is 1 foot-candle when the unit of length is a foot. For example, the amount of illumination on 1 ft^2 of area of a sphere of 1 ft radius is 1 lumen from a 1-candle source at center of sphere, and the intensity of illumination at any point is 1 ft-candle. A 1-candle source emitting light waves in all directions emits 4π lumens ($4\pi r^2$ = surface area of sphere). For a source of I candles, the light flux is $F = 4\pi I$ lumens.

A surface of A ft^2 receiving a total flux of F lumens has an average intensity of illumination of $E = F/A$ lumens per ft^2. The intensity of illumination E from a point source varies inversely as the square of the distance from the source.

(a) $$E = \frac{I}{r^2} \text{ ft-candles (or lumens/ft}^2\text{)}$$

The source intensity is I candles.

The desired or required amount of illumination is based on experience. For example, a recommended value for classrooms is 50 ft-candles. If the room is 25 feet $\times$ 30 feet = 750 ft^2 in area, then the light sources should be designed to deliver (750)(50) = 37,500 lumens to the working area. Tables that give the outputs in lumens of various kinds of lamps and fixtures are available. A *coefficient of utilization* must also be applied to account for the percentage of the total output of the light that reaches the working surface. (This varies with the light fixture and wall colors. Some light is absorbed by other than the working surface.)

REFLECTION AND REFRACTION

When a beam of light (or other electromagnetic emanation) strikes a body, it can be (1) reflected, (2) absorbed, (3) transmitted through the body, or (4) some combination of the three phenomena.

The angle of incidence (between ray and normal) is equal to the angle of reflection. Reflection from a rough surface is diffused.

Spherical Mirror

In figure 14.1, the center of the *concave* spherical mirror is C, its radius is r, the forcus is F, and the focal length is f. All rays of light from point A, such as AE and AG, are reflected through point A'. Point A' is the image of A. $A'B'$ is the image of AB. q is the *image distance.* The relation between the various dimensions is

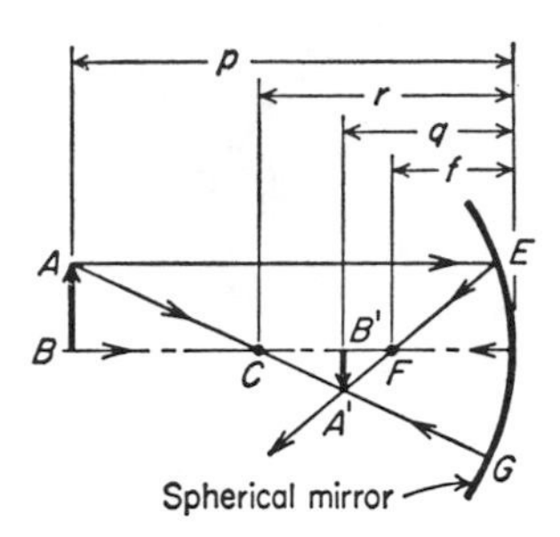

Figure 14.1

(b)
$$\frac{1}{p} + \frac{1}{q} = \frac{2}{r} = \frac{1}{f}$$

This is closely true when the curvature of the mirror is slight so that the tangent of the angle of reflection is equal to the angle. Similar statements can be made about a *convex* spherical mirror, shown in figure 14.2, and equation (b) holds except that q and f must be substituted as negative numbers, since they are measured in the opposite direction from p. Magnification is given by the ratio q/p.

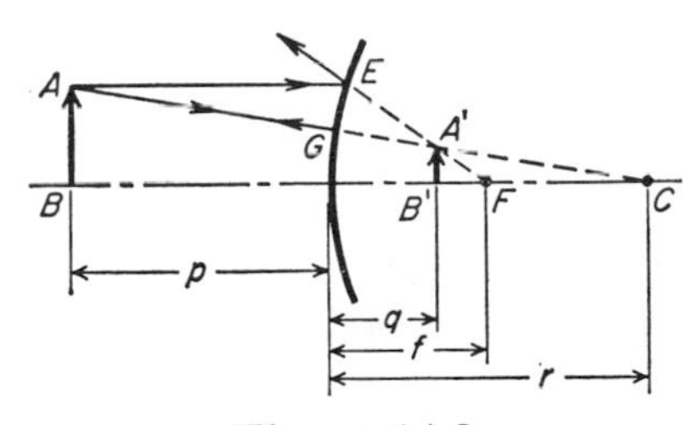

Figure 14.2

Refraction

If a ray of light (or other wave) passes into another medium obliquely, it is deviated toward the normal to the surface when its speed in the new medium is smaller ($v_2 < v_1$) and away from the normal when it changes to a greater speed ($v_2 > v_1$). This phenomenon is called *refraction.* The *refractive index*, shown in figure 14.3, is

(c)
$$\mu = \frac{v_1}{v_2} = \frac{\sin(\text{incidence angle})}{\sin(\text{refractive angle})} = \frac{\sin i}{\sin r}$$

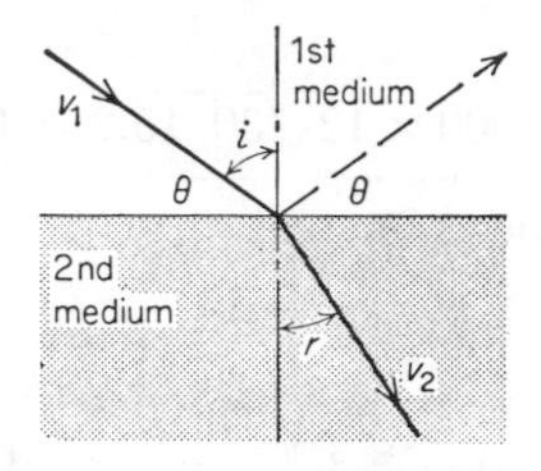

Figure 14.3

v_1 is the speed of the ray in the first medium and v_2 is its speed in the second medium into which it passes. Tabular values are generally for a vacuum to medium 2, virtually the same as air to medium 2. It also varies somewhat with wavelength, which accounts for the fact that white light passing through a prism disperses into a spectrum.

Apparent Depth

An observer in medium 1 looking at an object submerged in medium 2 sees the object at an apparent depth such that

(d) $$\text{actual depth} = \mu(\text{apparent depth})$$

Depth is as measured from the surface of medium 2, where μ is the refractive index from medium 1 to medium 2. If μ is greater than 1, the apparent depth is less than the actual.

SOUND

In solids, the speed of sound is given by the same equation as for liquids, with E defined as the modulus of elasticity.

(e) $$v = \sqrt{\frac{E}{\rho}}$$

ρ is the density. For example, steel has $E = 3.0$ EE7 psi, where the pound is a pound force, and $\rho = 490$ lbm/cu ft. If the pound is used for force (the usual choice when a choice is necessary), then the mass must be in slugs for consistency:

$$\rho = 490/32.2 = 15.2 \text{ slugs/cu ft}$$

$$E = 3.0 \text{ EE7} \times 144 \text{ psf (foot unit used)}$$

Then,

$$v = 1000 \times 12\sqrt{30/15.2} = 16{,}900 \text{ ft/sec}$$

This is the acoustic speed in steel.

Sound from Vibrating String

For a sound wave produced by a vibrating body, T is the period of a vibration, $f = 1/T$ is the frequency (usually in Hz), and λ is the wavelength (distance between the crests of the sinusoidal curve that represents the wave). The wave moves a distance λ in time T and therefore has a speed of

(f)
$$v = \frac{\lambda}{T} = f\lambda$$
$$\lambda = \frac{v}{f}$$

Speed is determined by the properties of the medium as previously noted. A vibrating string produces a first harmonic wave whose length, λ, is $2L$ (as in figure 14.4), where L is the length of the string. It also produces harmonics whose wavelengths are $\frac{1}{2}$, $\frac{1}{3}$, $\frac{1}{4}$, and so forth, of $2L$. If the tensile force on the string is F and its mass per unit length is m/L, the speed of the wave is also

(g)
$$v = \sqrt{\frac{F}{m/L}}$$

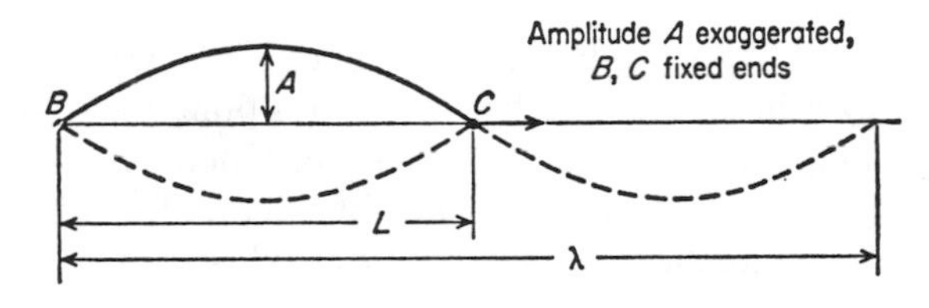

Figure 14.4

A consistent system of units must be used, say pounds and slugs. Combining equations (f) and (g), and using $\lambda = 2L$, we get the frequency of the first harmonic as

(h)
$$f = \frac{1}{2L}\sqrt{\frac{F}{m/L}}$$

The frequencies of the harmonics are n times the val·:e in (h), where n is an integer.

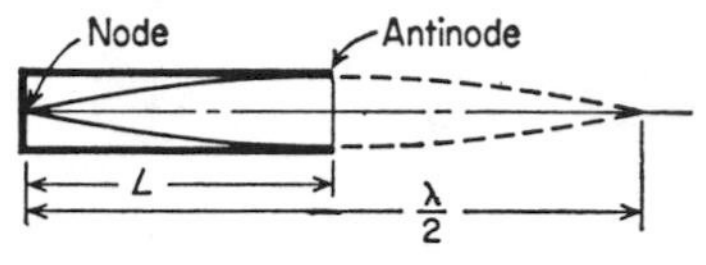

Figure 14.5

Sound from Gas Columns

The length of closed tube is a quarter of the wavelength of the first harmonic: $L = \lambda/4$ or $\lambda = 4L$. See figure 14.5. The length of tube open at both ends is half the length of the first harmonic: $\lambda = 2L$. The frequency from (f) is

(i)
$$f = \frac{v}{\lambda}$$

$$f = \frac{v}{4L} \quad \text{[CLOSED]}$$

$$f = \frac{v}{2L} \quad \text{[OPEN]}$$

v is the speed of sound in the gas in the column. The frequencies of the harmonics in closed-end tubes is n times the value of (i), where $n = 1, 3, 5, 7, \ldots$.

Beats

Waves of opposing phases produce interference. If two waves have different frequencies, they reinforce each other at certain points and oppose (interfere) at other points. If the difference in frequencies is small, two such sound waves produce beats, the frequency of the audible reinforcement. If the difference in the frequencies is d, one hears d beats per second.

PRACTICE PROBLEMS

14.1. A pump lifts 50 gpm of gasoline into a tank 60 feet above the intake of the pump. Specific gravity of the gasoline is 0.64. The pump is 65% efficient and is operated by an electric motor 90% efficient. What current in amperes would be required if the circuit is 210 volts, single phase?

Ans. 2.95 amps

14.2. A glass rod of a refractive index of 1.5 has hemispherical surfaces of 5 cm radius at each end. An object 25 cm from one end and outside of the rod creates a final image 30 cm from the opposite end but within the rod. What is the length of the rod?

Ans. 36.25 cm

14.3. A tuning fork of 512 frequency makes 2 beats/second with a vibrating string. Tightening the string eliminates the beats so that the fork and the string vibrate in unison. Find in what proportion the tension in the string was increased.

Ans. $(512/510)^2$

14.4. A fixed amount of air is confined at an unknown volume V and a pressure equivalent to 75 cm Hg. When the air is allowed to expand into an additional space of 40 cu cm, the pressure becomes 50 cm Hg. What is the original volume?

Ans. 80 cu cm

14.5. It is desired to heat 400 gallons of water from 60 °F to 100 °F by adding to it water that is at 200 °F. Calculate the number of pounds of 200 °F water necessary.

Ans. 1333 lbm

14.6. A piece of lead (specific gravity 11.3) is attached to 40 cc of cork (specific gravity 0.25). When fully submerged, the combination will just float. What is the weight of the lead?

Ans. 32.9 g

14.7. A 20 candle-power standard lamp and a lamp to be tested are placed 200 cm apart. It is found that the two lamps produce equal illumination on a screen placed between them when the screen is 80 cm from the standard lamp. What is the candle power of the lamp under test?

Ans. 45 c

14.8. A cast-iron sphere is hollow. It weighs 8.72 lbm in air and 6 lbm in water. The specific gravity of cast iron is 7.42. Assuming that the center void is spherical, find the volume of the void.

Ans. 11.48 in^3

14.9. A tuning fork making 256 vibrations/second is in resonance with a tube of hydrogen 4 feet long. Find the velocity of sound in the hydrogen.

Ans. 4100 ft/sec

14.10. An empty bottle weighs 50 grams. The same bottle full of water weighs 200 grams. Some dry sand is put into the empty bottle, and the sand and bottle together weigh 320 grams. The bottle containing the sand is now filled with water and the bottle, sand, and water together weigh 370 grams. Assuming that water weighs 1 g/cc, find the specific gravity of the dry sand.

Ans. 2.7 g/cc

14.11. What is the frequency of a violin string that gives out waves 24 inches long? How long will it take one of these waves to reach the back of the concert hall, a distance of 220 feet? The velocity of sound in air is 1130 ft/sec.

Ans. 565 Hz, 0.195 sec

14.12. The density of steel is 490 lbm/cu ft and Young's modulus for steel is 2.9 EE7 psi. What is the speed of sound in ft/sec in steel rails?

Ans. 16,600

14.13. In a scientific investigation, an airplane attained an altitude where the atmospheric pressure was 13.73 inches Hg and the temperature was −40 °F. If the weight-density of air is 0.081 lbm/cu ft at 29.92 inches Hg and 32 °F, what was the density of the air at the altitude reached by the airplane, assuming the usual gas laws apply?

Ans. 0.0435 lbm/cu ft

14.14. Two 6-volt batteries, each with an internal resistance of 0.05 ohm, are connected in parallel to a load resistance of 9.0 ohm. How much current flows through the load?

Ans. 0.665 amps

14.15. The upper edge of a vertical gate in a dam lies along the water surface. The gate is 6 feet wide and is hinged along the bottom edge, which is 10 feet below the water surface. What is the torque about the hinge?

Ans. 62,400 ft-lbf

15 *ENGINEERING ECONOMY*

SYMBOLS

C = cost
D = deposit, investment; depreciation
I = total interest, dollars
i = interest rate as a decimal
J = salvage (junk) value
n = number of interest periods, or number of years
P = principal dollars
S = value of a principal or deposit with interest

All examinations have some questions concerning the monetary aspects of a simple engineering proposal, often related to comparative returns from possible alternative actions. Such problems have to do with interest, depreciation, and other cost factors.

INTEREST

Simple Interest

Simple interest is used only for relatively short periods of time or where interest due is paid at short intervals (for example, monthly or semiannually). If you are to pay i interest per annum for n years on a principal sum of P, the interest is

$$I = Pin$$

The total amount S to be repaid is

$$S = P + I = P(1 + in)$$

Compound Interest

Interest is said to be compounded when the interest due is added to the principal P, thus forming a new principal for which interest is computed and added again at the end of the next interest period. Savings accounts generally operate on this basis. The equation is

(a) $$S = P(1 + fi)^{n/f}$$

S is the amount at the end of n years at the annual rate of interest of i on principal P, compounded at periods f when f is the fraction (usually) of a year (for example, for interest compounded annually, $f = 1$; for interest compounded semiannually, $f = \frac{1}{2}$; and so forth).

EXAMPLE 1: Calculate the amount that must be invested at 4% compounded annually to amount to \$500 in 10 years.

SOLUTION:

$$500 = P(1 + 0.04)^{10}$$
$$P = \$337.83$$

EXAMPLE 2: Calculate the amount when \$337.83 is invested at 4% compounded quarterly for 10 years.

SOLUTION:

$$S = 337.83\left(1 + \frac{0.04}{4}\right)^{\frac{10}{1/4}} = 337.83(1.01)^{40} = \$503.00$$

Sinking Funds

A sinking fund is a fund being built up to a particular amount at the end of a specified period by regular periodical deposits that draw compound interest. Sinking funds are often set up, actually or on paper, to provide money to pay off an indebtedness at a certain time. If the deposits D to the sinking fund account are made *annually*, the amount S at the end of n years is

(b) $$S = D\left[\frac{(1+i)^n - 1}{i}\right]$$

i is the annual interest rate, compounded. Tables for compound-interest and sinking-fund problems are available.

EXAMPLE: Calculate the sum that must be set aside annually in order to accumulate \$30,000 to replace a certain machine after 20 years, at 6% compounded. Salvage value of the machine is expected to be \$600.

SOLUTION: The amount to be accumulated is

$$S = 30{,}000 - 600 = \$29{,}400$$

$$29{,}400 = D\left[\frac{(1+0.06)^{20} - 1}{0.06}\right]$$

$$D = \$799.68 \quad (\textit{Ans.})$$

DEPRECIATION

In a business sense, depreciation is that part of the purchase price of capital goods that is counted as current cost of operating the business. Sometimes, depreciation is a legal bookkeeping charge for tax purposes. The amount of the charge during any one year can be decided in various ways. Two methods are described here.

Straight-Line Method

The life of the equipment is estimated, as well as its salvage value as junk or on the second-hand market at the end of the period. Then the total depreciation, which is original cost C minus final market value J, is divided by the expected years of life n to obtain a constant annual depreciation figure D:

(c) $$D = \frac{C - J}{n}$$

Sinking-Fund Method

The amount to be depreciated is $C - J$, as before, and it is to be accumulated by actual investment or a bookkeeping operation in accordance with the principal of sinking funds. Converting equation (b) to this purpose, the annual depreciation charge is

$$D = \frac{(C - J)i}{(1+i)^n - 1} \quad \text{(d)}$$

EXAMPLE: An asset that cost $22,000 is expected to have a salvage value of $2000 at the end of 20 years. What is the annual depreciation charge by the sinking-fund method if the interest rate is 4%?

SOLUTION:

$$D = \frac{(C - J)i}{(1+i)^n - 1} = \frac{(20{,}000)(0.04)}{1.04^{20} - 1} = \$672 \quad (\textit{Ans.})$$

PRODUCTION COSTS

The selling price of a product consists of the following elements:

1. **Direct material cost**. This is the cost of the raw material that goes into the product.

2. **Direct labor cost**. This is labor that can be allocated directly to the cost of manufacturing the product.

3. **Factory expense or overhead**. This includes depreciation, repairs, indirect labor (as janitor work), heat, light, power, insurance, taxes, and supervisory and clerical workers in the manufacturing division.

   Cost 1 + Cost 2 is called *prime cost.*

   Cost 1 + Cost 2 + Cost 3 is called *factory cost.*

4. **Administrative expense**.

5. **Selling expense**.

6. **Profit (or loss).**

Costs $(1 + 2 + 3 + 4)$ is called *production cost.*
Costs $(1 + 2 + 3 + 4 + 5)$ is called *total cost.*

PRACTICE PROBLEMS

15.1. A manufacturing company buys a machine for $50,000. It estimates that the machine can be used 20 years before it is worn out and that it can then be sold for $10,000. What should be the yearly charge against the machine at 6% interest? Figure the charge by the straight-line depreciation method.

15.2. A manufacturer is planning to produce a new line of products that will require buying or renting new machinery. A new machine will cost $17,000 and have an estimated value of $14,000 at the end of 5 years. Special tools for the new machine will cost $5000 and have an estimated value of $2500 at the end of 5 years. Maintenance costs for the machine and tools are estimated to be $200 per year. What will be the average annual cost of ownership during the next 5 years if interest is 6%? Assume that the annual cost is equal to the straight-line depreciation.

15.3. An old light-capacity highway bridge can be strengthened at a cost of $9000, or it can be replaced by a new bridge of sufficient capacity at a cost of $40,000. The present net salvage value of the old bridge is $13,000. It is estimated that the old bridge, when reinforced, will last for 20 years, with a maintenance cost of $500 per year and a salvage value of $10,000 at the end of 20 years. The estimated salvage value of the new bridge after 20 years of service is $15,000. The maintenance on the new bridge will be $100 per year. If the interest is 6%, determine whether it is more economical to reinforce the old bridge or to replace it.

15.4. The owners of a concrete batching plant use a power steam shovel which, if repaired at a cost of $2000, would last another 10 years. Maintenance costs, operating costs, taxes, and insurance would be $2500 per year. The present salvage value of the power steam shovel is $6000 and, if repaired and used for 10 years, the estimated salvage value will be $500. A new power shovel could be purchased at a cost of $30,000, and its estimated salvage value at the end of 10 years would be $20,000. The maintenance costs, operating costs, taxes, and insurance would be $1500 per year. If the interest rate is 6%, determine whether it is more economical to repair and use the old steam shovel or buy the new one. Use straight-line depreciation.

15.5. An individual wishes to deposit a certain quantity of money so that at the end of 5 years, at 4% interest compounded semiannually, he will have $500. How much must he deposit?

Ans. $410.15

15.6. The purchase of a $15,000 machine is contemplated. However, it is believed that this is a highly inflated price, and that the machine will probably be salable for only $7000 if it is disposed of 5 years hence. If annual taxes are 2% of first cost, insurance is 0.5%, and upkeep is $250 per year, what is the prospective annual cost of ownership of this machine if it is to be disposed of in 5 years? Will the purchase of this machine be more economical than renting a similar one at $200 a month? Interest is $4\frac{1}{2}$%. Calculate annual cost by straight-line depreciation.

15.7. Money is invested at a nominal rate of interest of 5% per annum compounded semiannually. What is the effective rate per annum?

Ans. 5.06%

15.8. A lathe costs $10,000 now. It has a life expectancy of 20 years and an estimated salvage value of $2000 at the end of 20 years. At 7% interest, what is the annual "capital recovery with a return" cost of the lathe on the basis of the above estimates? Use straight-line depreciation in your calculations.

15.9. A steam power plant costs $1,000,000 now. It has a life expectancy of 40 years and its estimated salvage value is $100,000. The operating cost of the plant is $30,000 per year. At 6% rate of interest, determine (a) the annual straight-line depreciation, (b) the annual average interest on the investment, (c) the total annual cost.

Ans. (a) $22,500, (b) $33,675, (c) $86,175

15.10. A manufacturing plant is considering the purchase of additional equipment for producing a machine part for the aircraft industry. Machine *A*, with an initial cost of $12,000 and a life of 15 years, will produce 20,000 units per year with a monthly maintenance cost of $50. Machine *B* produces the same part by stamping at the same yearly rate with an initial cost of $15,000 and a life of 30 years. However, the die assembly for stamping must be replaced every 10 years at a cost of $5000 and a salvage value of $2000. The yearly maintenance cost for machine *B* will be $60. With interest in all cases at 6% and using straight-line depreciation, which machine would you purchase if the salvage value of each machine is considered to be zero at the end of their respective lives?

15.11. Money is invested at 4% per annum compounded quarterly. What is the annual rate of interest?

Ans. 4.06%

15.12. The quantity (Q gallons) of a certain mineral water that can be sold at various prices (P cents per gallon) is shown in the following table. The total expense ($\$E$) of marketing these quantities is also shown.

P cents/gallon	Q gallons	$\$E$
20	16,200	2,700
30	12,800	2,400
40	9,800	2,100
50	7,200	1,800
60	5,000	1,500
70	3,200	1,200

What price per gallon yields the greatest net profit?

Ans. 40 ¢/gallon

15.13. A certain type of machine A has an estimated life of 10 years. Another type of machine B will perform the same service as A and will last considerably longer. The initial investment in B will be 40% greater than that in A. Operation and maintenance costs in both cases are very small and should be neglected. If the minimum desired return on the money invested is 6%, what must be the minimum expected life of B to make it a good investment?

Ans. 16 years

15.14. A certain untreated structure has a first cost of $15,000 and an estimated life of 12 years. It is expected that treatment with a certain timber preservative will increase the prospective life to 20 years and reduce the average annual maintenance costs by $300. If interest is figured at 8%, what is the maximum amount it is justifiable to pay for timber treatment?

Ans. $8920

15.15. A telephone company finds there is a net profit of $15.00 per instrument if an exchange has 1000 or fewer subscribers. If there are over 1000 subscribers, the profit per instrument decreases 1 cent for each subscriber above that number. How many subscribers would give the maximum net profit?

Ans. 1750

15.16. Prepare a table to determine the economical lot size of a manufactured product under the following conditions: Preparation cost is $400 per lot; annual demand is 500,000 pieces; the plant works 300 days per year; daily production when operating is 3000 pieces; material, labor, and manufacturing overhead is 5 cents per piece; and minimum return desired on investment of working capital is 20%. Prepare the table using lot sizes that are multiples of 10,000 pieces per lot.

15.17. An asset has a first cost of $13,000.00, an estimated life of 15 years, and a salvage of $1000.00. For depreciation, use the sinking-fund method with interest at 5% compounded annually. (a) Find the annual sinking-fund annuity or depreciation charge. (b) Find the balance in the sinking fund (i.e., the amount accumulated toward depreciation of the asset) at the end of 9 years. (c) If the asset were to be sold for $4000.00 at the end of 9 years, what would be the net *book value gain* or *loss*?

Ans. (a) $555.60, (b) $6128, (c) $2872 loss

15.18. A certain product is selling for $8.20. The production cost is $7.38. (a) What percentage of the selling price represents profit? (b) What percentage reduction in production costs will increase the margin of profit 60%?

Ans. (a) 10%, (b) 6.6%

15.19. One year ago, a special-purpose machine was purchased for a certain operation at a cost of $2000. It was estimated the machine would be good for 5 years. Because of a faulty design, the machine has failed to function properly, making it necessary to shut down the machine at frequent intervals for repairs and adjustments. The cost of repairs and adjustments has been $200 per month. A new, well-designed machine which does the same job is now obtainable at a cost of $3500. The cost for repairs and adjustments has been estimated at $50 per month. The new machine should last 5 years. The old machine has no salvage value. Other than repairs and adjustments, the operating costs for the two machines are substantially equal. Use 8% interest. Should the new machine be purchased?

Ans. Yes

15.20. A loan company advertises that $100.00 borrowed for one year may be repaid by 12 monthly payments of $9.46. Assuming the difference between the amount repaid and the amount borrowed is interest only, what is the effective annual interest rate?

Ans. 26.8%

INDEX

V

W

Y

Z